Low-Temperature Selective Catalytic Reduction Catalysts

Weiguo Pan · Ruitang Guo

Low-Temperature Selective Catalytic Reduction Catalysts

Science Press
Beijing

Springer

Weiguo Pan
College of Energy and Mechanical Engineering
Shanghai University of Electric Power
Shanghai, China

Ruitang Guo
College of Energy and Mechanical Engineering
Shanghai University of Electric Power
Shanghai, China

ISBN 978-981-97-9297-9 ISBN 978-981-97-9298-6 (eBook)
https://doi.org/10.1007/978-981-97-9298-6

Jointly published with Science Press
The print edition is not for sale in China Mainland. Customers from China Mainland please order the print book from: Science Press.

This Springer imprint is published by the registered company Springer Nature Singapore Pte Ltd.
The registered company address is: 152 Beach Road, #21-01/04 Gateway East, Singapore 189721, Singapore

Contents

Chapter 1
The Harm of NO_x and Its Emission

1.1 NO_x

1.1.1 The Characteristic of NO_x

Nitrogen oxide (NO_x) refers to a compound composed only of nitrogen and oxygen, including a variety of compounds, such as nitrogen monoxide (N_2O), nitric oxide (NO), nitrogen dioxide (NO_2), nitrogen trioxide (N_2O_3), nitrogen tetroxide (N_2O_4) and nitrogen pentoxide (N_2O_5). In addition, there are nitrosyl azide (N_4O) and nitrogen trioxide (NO_3). Trinitroamine $N(NO_2)_3$ is also a compound composed only of nitrogen and oxygen elements, but it is not an oxide in the strict sense. Except for nitrous oxide and nitrogen dioxide, other nitrogen oxides are unstable. When exposed to light, humidity, or heat, they become nitrogen dioxide and nitric oxide, and nitric oxide becomes nitrogen dioxide. Therefore, the environment is exposed to several gas mixtures, often called nitrous smoke (gas), mainly nitric oxide and nitrogen dioxide, and mainly nitrogen dioxide. Nitrogen oxides have different degrees of toxicity.

N_2O_3 and N_2O_5 are acidic oxides. The corresponding acid of N_2O_3 is nitrous acid (HNO_2), and N_2O_3 is the anhydride of nitrous acid; The corresponding acid of N_2O_5 is nitric acid, and N_2O_5 is the anhydride of nitric acid. NO, N_2O, N_2O_4 and NO_2 are not acid oxides. Nitrogen tetroxide (N_2O_4) is a dimer of nitrogen dioxide (NO_2), which is often mixed with nitrogen dioxide (NO_2) to form an equilibrium mixture. The mixture of nitric oxide (NO) and nitrogen dioxide (NO_2), also known as nitrate gas (nitrate smoke). They are slightly soluble in water, and the aqueous solution is acidic in varying degrees. Nitric oxide (NO) and nitrogen dioxide (NO_2) decompose into nitric acid and nitrogen oxide in water. Nitrous oxide (N_2O) has strong oxidation only when it is above 300 °C, and the rest have different degrees of oxidation, especially nitrogen pentoxide (N_2O_5), which decomposes above − 10 °C to release oxygen and laughing gas. Nitrogen oxides are noncombustible substances, but they can support combustion. For example, nitrogen monoxide (N_2O), nitrogen

W. Pan and R. Guo, *Low-Temperature Selective Catalytic Reduction Catalysts*,
https://doi.org/10.1007/978-981-97-9298-6_1

dioxide (NO_2) and nitrogen pentoxide (N_2O_5) can cause explosion in case of high temperature or combustible substances.

Under the condition of high-temperature combustion, NO_x mainly exists in the form of NO, and NO accounts for about 95% of the initial NO_x emission. However, NO easily reacts with oxygen in the air to generate NO_2 in the atmosphere, so NO_x generally exists in the form of NO_2 in the atmosphere. NO and NO_2 in the air reach equilibrium through photochemical reaction and mutual conversion. When the temperature is high or there are clouds, NO_2 further interacts with water molecules to form nitric acid (HNO_3), the second important acid in acid rain. In the presence of catalyst, if appropriate meteorological conditions are added, the conversion rate of NO_2 to nitric acid is accelerated. Especially when NO_2 and SO_2 exist at the same time, they can catalyze each other and form nitric acid faster.

1.1.2 The Harm of NO_x

The atmospheric chemistry of nitrogen oxides has indirect consequences on radiative forcing, as the direct greenhouse effect through the absorption of longwave radiation of trace gases NO and NO_2 is negligible.

The first indirect greenhouse effect is that the NO_x produces tropospheric ozone. Depending on the concentration of the NO_x level, there is either a formation or loss of ozone (O_3) in troposphere. If the NO_x less a particular threshold value, ozone is removed, while exceeding this valve it is formed, as shown in Fig. 1.1. The NO_x concentration increases in regions of the troposphere which are low in NO_x (primarily the regions in the tropics), resulting in an alteration from a net O_3 loss to a net O_3 production. In regions of the troposphere which are rich in nitrogen oxides, additional introduction of NO_x causes the ozone concentration to become even higher. A molecule of NO_2 formed near the ground in a rural, anthropogenically influenced region of the northern hemisphere leads to the production of up to 12 molecules of O_3 (experimental findings by [1–4]). We know that ozone is a greenhouse gas, so the NO_x is one of global warming potentials (GWP) gases.

The second indirect greenhouse effect is that the NO_x controls the atmospheric concentration of the hydroxyl radical which is the most important species of tropospheric chemistry. The hydroxyl radical mediates the removal of the most trace substances including that of methane (CH_4). The increase of NO_x concentration leads to the decrease of hydroxyl concentration, which increases the concentration of methane and other trace gases. This is another reason for NO_x as a greenhouse gas.

Nitrogen oxides are converted to nitric acid and particulate nitrate (aerosol) in the atmosphere. This occurs within only a few days. In the end nitrogen oxides are converted to nitrous oxide by the soils. And N_2O has a long atmospheric residence time and is recognized as a strong greenhouse gas. Nitrous oxide is a powerful greenhouse gas, though its concentration is still very small relative to that of other major greenhouse gas. There are two attributes for this result. First, it is chemically

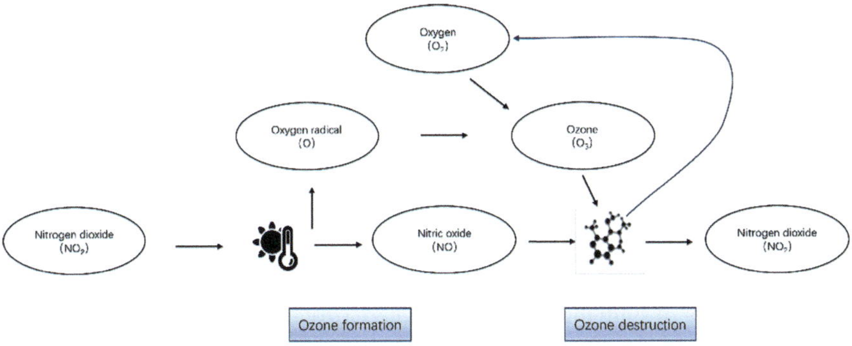

Fig. 1.1 Nitrogen oxides and ozone [5]

rather inert, which means that once it is emitted into the atmosphere it can remain there for many decades—the average molecule of nitrous oxide in our atmosphere has a lifetime of around 120 years. Second, every nitrous oxide molecule is itself a very effective absorber and re-emitter of heat (infrared radiation). Infrared radiation is emitted from Earth at a range of wavelengths from 0.7 to 70 microns in size. Across much of this range the outgoing energy is absorbed and re-emitted by water vapour—the leading greenhouse gas on our planet—while at some wavelengths carbon dioxide and methane are the key greenhouse gases. It is at wavelengths of 4–5 and 7–8 microns that nitrous oxide becomes a big player. In these narrow atmospheric windows that would otherwise allow heat radiation to escape from Earth, nitrous oxide effectively blocks the way—making it an especially powerful greenhouse gas. These properties of long life and strong infrared absorption mean that, over a 100-year time horizon, each ton of nitrous oxide emitted has a global warming potential (GWP) equivalent to releasing 300 tons of carbon dioxide [5].

As we have already seen, in the lower atmosphere (called the troposphere) nitrous oxide is such a powerful greenhouse gas because it is so stable and lasts for many decades. It is when nitrous oxide molecules seep up into the upper atmosphere (the stratosphere) that they are prone to attack by a combination of sunlight and oxygen radicals. Most of the nitrous oxide is destroyed by sunlight, but where it instead reacts with oxygen radicals it forms nitrogen oxides, and it is these highly reactive substances that can then go on to deplete ozone. As we know, there have been many photochemical smog incidents in history, especially in Los Angeles in 1970, causing 75% of the population to suffer from red eye disease. In 1971, a serious photochemical smog event occurred in Tokyo, Japan, which caused poisoning and coma to some students. Many photochemical smog events have occurred in China with the industrialization process. As we know, nitrogen oxides are the main cause of photochemical smog.

Nitrogen dioxide is toxic to humans, and if its concentration in the air tops 150 parts per million it will begin to kill those who breathe it. Death can be rapid, with the lungs swelling and filling with fluid, or may take several weeks as the fine branches of the lungs—the bronchioles—become so inflamed that air supply is shut off and the victim slowly suffocates. Even at much lower levels, chronic exposure to nitrogen dioxide can cause serious damage to the lungs, and its presence has been implicated in cases of coronary heart disease and sudden infant death syndrome. The direct effects of more NO_x can therefore be severe, and efforts to reduce its emissions have intensified in the past few decades. But, for the health department of almost every major city in the world more NO_x in the air heralds something even more dangerous ozone. Ozone, that precious high-altitude gas that shields our planet and everything that lives on it from the most damaging rays of the sun, is also formed closer to earth by an excitable combination of NO_x gases, other air pollutants and the crucial active ingredient of sunlight (collectively called 'ozone precursors'). It is this potent mixture that helps give rise to the modern-day summertime smog now endured by so many large urban areas. Often called 'photochemical smog', ozone is its most infamous constituent and can cause severe internal burning and respiratory damage if breathed in. As the amounts of ozone in the air begin to increase, so the damage to the lungs of those breathing it becomes more severe: at just one part in every million the passages of the lungs become irritated and inflamed; at two parts per million even short periods of exposure can cause severe lung damage and death.

A further potential human health impact of NO_x comes via its conversion into small particles of airborne nitrate through reaction with nitric acid. The risks from such particulate matter (PM) largely depend on the size of the particles, with the smallest being the most dangerous. For PM containing nitrate, little evidence exists on whether these particles pose a significant threat to human health, but it is known that they can contribute to the mix of fine particles in the air we breathe.

1.2 The Formation Mechanism of NO_x

1.2.1 NO_x Formation Mechanism and the Interfering Factors

In general, nitrogen oxides (NO_x) can be formed from oxidation of nitrogen compounds in the fuel (fuel-NO_x) and the fixation of atmospheric nitrogen (thermal-NO_x). In addition, there is also a part of molecular nitrogen in the early stage of the flame, which is converted into NO_x through intermediate products, under the influence of nitrogen hydrocarbon compounds. It is called prompt NO_x. This part is very small and generally not considered.

Nitrogen dioxide is also formed in combustors. However, the route to NO_2 is through NO. Therefore, NO is sufficient for determining NO_x. And most of the NO_x emitted by combustion sources is NO with only a small fraction (typically 5% or less) appearing as NO_2.

The formation mechanism of thermal NO_x is complicated. Generally thermal NO is calculated via the extended Zeldovich mechanism.

$$O + N_2 \rightarrow NO + N \tag{1.1}$$

$$N + O_2 \rightarrow NO + O \tag{1.2}$$

$$N + OH \rightarrow NO + H \tag{1.3}$$

The first step (reaction 1.1) is rate-limiting and has a very high activation energy (75 kcal/gmol). This high activation energy implies that high temperatures are required for significant NO formation, hence the designation "thermal" NO [6].

As shown by Bowman [7], the following expression is obtainable

$$\frac{d(NO)}{d(t)} = \frac{1.44 \times 10^{20}}{T^{1/2}} \exp\left(\frac{-69500}{T}\right)(O_2)^{\frac{1}{2}}(N_2),\ \text{Kgmol/m}^3/\text{s} \tag{1.4}$$

where T is gas temperature in Kelvins, t is reaction time in seconds, and all gas concentrations are in units of Kgmol/m^3. This rate expression shows that thermal NO formation is highly temperature dependent, and is also dependent on oxygen concentration and residence time. Temperature effects are particularly pronounced. If the temperature is quite low and the oxygen atom number of free radicals is not enough, the formation of thermal NO_x will be suppressed.

Fuel NO_x is formed during combustion due to the oxidation of nitrogen contained in the fuel. The formation mechanism of fuel NO_x is more complicated than thermal NO_x, and an exact determination of complete mechanism is presently not available. There are hundreds of reactions involved, and many difficult-to-measure intermediates and radical species.

In generally, the fuel NO_x mechanism includes a rapid conversion of fuel nitrogen compounds into intermediate nitrogen compounds (HCN, CN, NH_2, NH, N) which can either be converted to NO by attack of oxygen containing species or be converted to N_2 by reaction with NO itself [7–10]. The reaction pathway can be shown in the Fig. 1.2.

Factors affecting fuel NO_x formation are the composition of nitrogen in fuel and the concentration of oxygen in the gas. Although fuel NO_x cannot be correlated with nitrogen content alone, increased fuel nitrogen content can lead to higher emissions of fuel NO_x. In general, the higher the oxygen concentration in the gas is, the more fuel NO_x forms. Fuel NO_x forms even at low temperature.

In summary, the main factors affecting NO_x formation in fuel combustion chamber are as follows:

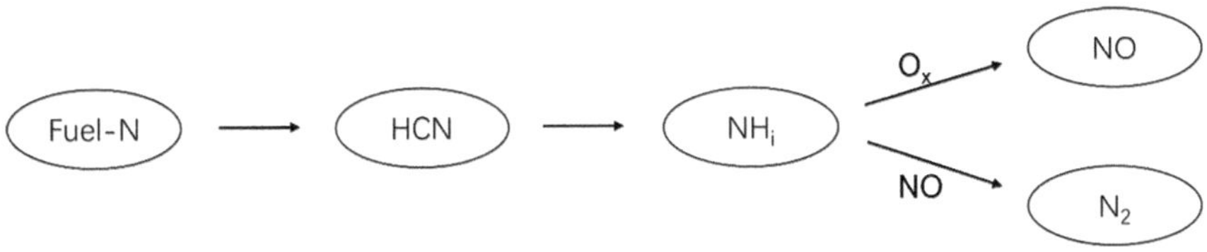

Fig. 1.2 The pathway of fuel NO_x formation

(1) Fuel nitrogen content. The higher the nitrogen compound content, the more fuel NO_x generated. For example, the content of nitrogen compounds in gas fuels is very small, so NO_x generated during combustion is almost converted from nitrogen in the air. On the contrary, when burning solid fuel coal, especially pulverized coal, most of the nitrogen oxides in the flue gas (90%) are converted from the inherent nitrogen compounds in the fuel. Liquid fuel is between the above two.
(2) Flame temperature and residence time. The higher the temperature, the easier NO_x formation, especially thermal NO_x. When the temperature is above 2000 °C, it can be oxidized almost instantaneously to NO. In the range of 1600–2000 °C, if the residence time is longer, NO_x is also easy to be generated. If the duration is shorter, the NO_x generation rate is slower. When the temperature is below 1500 °C, the formation rate of thermal NO_x is significantly slower, but the formation rate of fuel NO_x remains unchanged.
(3) Oxygen concentration in combustion zone. With the increase of oxygen concentration in the combustion zone, the production of both thermal NO_x and fuel NO_x increases. In addition, when the oxygen supply is moderate and the combustion temperature is high, NO_x is more likely to be generated.

Among the above factors, flame temperature has a great influence on NO_x formation. The higher the temperature, the more NO_x generated. In addition, the formation of NO_x is also closely related to the combustion mode and the form of combustion device.

1.2.2 *Sources of NO_x*

As human impacts on the global nitrogen cycle grew rapidly from the industrial revolution onwards, human-induced nitrous oxide emissions also increased. Today, so-called 'natural sources' still account for 60% of worldwide emissions each year, but the human-induced component is becoming an ever-larger part of the global budget.

Manmade NO_x emissions mainly come from energy consumption. With the development of economy, an increase in per capita income will increase pollutants emissions until the turning point, and pollutants emissions will present a negative growth as per capita income increases after the turning point. Emissions of nitrogen oxides in China peaked in 2011, with 24.043 million tons [11]. In 2020, total nitrogen oxides emission in China reached 10.197 million tons [11]. This is principally a product of industries and vehicles. Of these, industrial NO_x emissions are 4.175 million tons, accounting for 40.9% of national NO_x emissions; domestic NO_x emissions amount to 0.334 million tons, accounting for 3.3% of national NO_x emissions; the emission of nitrogen oxides from mobile sources was 5.669 million tons, accounting for 55.6% of the national emission of nitrogen oxides; centralized pollution control facilities emit 19,000 tons of nitrogen oxides, accounting for 0.2% of national emissions. The national and source NO_X emissions in 2016–2020 are shown in Tables 1.1 and 1.2.

From the tables, we know that industrial and domestic sources of nitrogen oxides have decreased almost annually since 2016. The power plant sector is regard as the first priority for NO_X emissions reduction in the Chinese national 12th five-year-plan (2011–2015). During this period, many power plants were transformed and nitrogen removal equipment installed in China. Accordingly, we can see that the industrial sources of nitrogen oxides have dramatically decreased since 2011. The other possible reason is that the government and citizens are becoming more environmentally conscious with the development of economy. Aside from this, the technological advances that accompany economic growth also reduce the emissions of pollutants.

In 2020, among the 42 industries surveyed, the top three industries ranked for NO_X emissions were electricity, thermal production and supply, non-metallic mineral products, black metal smelting, and rolling processing. The total nitrogen oxide emissions of the three industries are 3.288 million tons, accounting for 78.8% of the national industrial nitrogen oxide emissions. Figure 1.3 expresses NO_X emissions from industrial sources in 2020.

In 2016, NO_X emissions from thermal power plants amounted to 1.917 million tons, accounting for 23.7% of national industrial NO_X emissions. In 2020, the NO_X

Table 1.1 The national and source NO_X emissions in 2016–2020

Year	Total (million tons)	Industrial NO_X emissions (million tons)	Domestic NO_X emissions (million tons)	Mobile NO_X Sources (million tons)	Centralized pollution control facilities (million tons)
2016	15.033	8.091	0.616	6.316	0.01
2017	13.484	6.465	0.592	6.412	0.015
2018	12.884	5.887	0.531	6.446	0.02
2019	12.339	5.481	0.497	6.336	0.024
2020	10.197	4.175	0.334	5.669	0.019

Table 1.2 The national and source NO_X emissions proportion in 2016–2020

Year	Total (proportion%)	Industrial NO_X emissions (proportion%)	Domestic NO_X emissions (proportion%)	Mobile NO_X Sources (proportion%)	Centralized pollution control facilities (proportion%)
2016	100	53.8	4.1	42	0.1
2017	100	47.9	4.4	47.6	0.1
2018	100	45.7	4.1	50	0.2
2019	100	44.4	4.0	51.4	0.2
2020	100	40.9	3.3	55.6	0.2

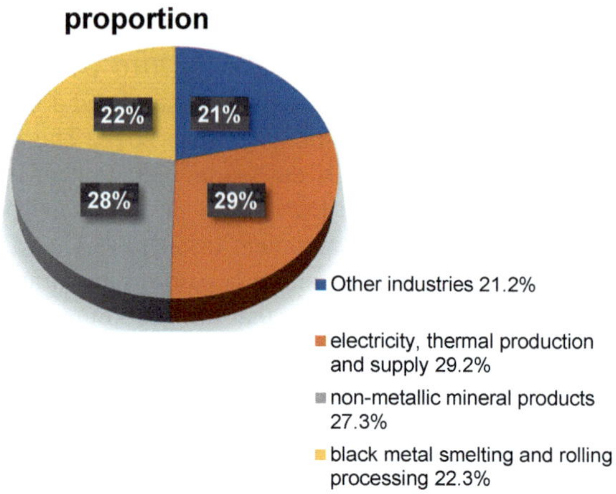

Fig. 1.3 The national and source NO_X emissions in China in 2020

emissions of thermal power enterprises were 0.612 million tons, accounting for 14.7% of the national industrial NO_X emissions. The main reason for this reduction is that thermal power plants are heavily installed with SCR or SNCR equipment.

Increased combustion of fossil fuels since the last century has been a primary source of NO_X, leading to the increase of pollutants concentration in the atmosphere. However, other sources of NO_X such as the production and use of nitric acid should not be neglected. During nitric acid plant operation as well as nitrification and oxidation of organic compounds with the use of nitric acid, nitrous gases in varying concentrations are formed [12].

Natural sources of NO_X, despite not being as crucial as anthropogenic ones, are still worth listing: oxidation of NH_3, lightning, and volcano activities.

References

1. Luke WT, Dickerson RR, Ryan WF, et al. Tropospheric chemistry over the Great Plains of the United States 2. Trace gas profiles and distributions. J Geophys Res 1992;97:20647–670.
2. Geiss H, Vole-Thomas A. Lokale und regionale Ozonproduktion: Chemie und Transport. Abschlut~bericht zum Vorhaben BMFT 07EU7235 (1993) (Report from the J/ilich research center No.Jiil-2764).
3. Kleinm L, Lee YN, Springston SR, et al. Ozone formation at a rural site in the southwestern United States. J Geophys Res. 1994;99:3469–82.
4. Olszyna KJ, Bailey KM, Simonaitis R, et al. Ozone and NO_X relationships at a rural site. J Geophys Res. 1994;99:14557–63.
5. Nitrogen and climate change an explosive story (Dave Reay (auth.)).
6. Nichols KM, Thompson LM, Empie HJ. A review NO_X formation mechanisms in recovery furnaces, Tappi J. 1991.
7. Bowman CT. Chemistry of gaseous pollutant formation and destruction, Chapter 4 of fossil fuel combustion: a source book. In: Bartok W, Sarofim AF, editors. New York, NY: John Wiley & Sons, Inc.; 1991.

8. Boardman RD. Development and evaluation of a combined thermal and fuel nitric oxide predictive model. Ph.D. Dissertation. Provo, UT: Chemical Engineering Department, Brigham Young University; 1990.
9. Fenimore CP. Reactions of fuel-nitrogen in rich flame gases. Combust Flame. 1976;26:249–56.
10. Soete GG. Overall reaction rates of NO and N_2 formation from fuel nitrogen. In: Fifteenth symposium (international) on combustion; 1975. pp. 1093–102.
11. Annual Report of Environmental Statistics 2011–2020 (China).
12. Dyer-Smith P, Jenny P. Application of ozone to avoid the production of nitrous gases (NO_x) during industrial nitrification and oxidations with nitric acid. In: IOA 17th world ozone congress—Strasbourg; 2005. VII.3.11.

Chapter 2
NO_x Emission Control Technologies (NO_x Emission Abatement)

2.1 Introduction

According to the formation mechanism of NO_x and sources of NO_x, what we can do about NO_x reduction is mainly about fossil fuel combustion. In the case of incineration processes occurring three methods to abatement of NO_x emission are known: precombustion, combustion modification, and post combustion techniques.

(1) Denitrification precombustion mainly converts fuel into low nitrogen fuel, which has high cost and less engineering application.
(2) Denitrification in combustion mainly refers to various combustion technologies to reduce NO_x. The cost of this method is low, but the denitrification efficiency is not high.
(3) Post-combustion methods mainly are dealing with nitrogen oxides in exhaust gases from incineration processes, which has high efficiency. With the increasingly stringent environmental requirements, high efficiency flue gas denitrification technology is the main development direction.

The following is a brief introduction and review of several denitrification technologies commonly used in coal-fired power plants.

2.1.1 Pre-combustion and Combustion Modification

Pre-combustion generally means fuels purification for the purpose of reducing the amount of nitrogen or choosing the fuel with low nitrogen content. Fuel type affects the formation of NO_x through the amount of fuel bound nitrogen [1]. Moreover, replacement of air by pure oxygen in the combustion process can also substantially decrease the formation of NO_x. Because neither thermal NO_x, nor prompt NO_x can be formed. The main drawback of this solution is its high cost.

W. Pan and R. Guo, *Low-Temperature Selective Catalytic Reduction Catalysts*,
https://doi.org/10.1007/978-981-97-9298-6_2

Combustion modification can be simply described as alteration of operational conditions in order to decrease the NO_x formation.

(1) Low excess air (LEA) combustion

The combustion process is carried out as close as possible to the theoretical air volume, and the formation of NO_x can be inhibited with the decrease of excess oxygen in flue gas. This is a simple way to reduce NO_x. In general, low excess air combustion can reduce NO_x emission by 15–20%. However, this method has some limitations, such as when the oxygen concentration in the furnace is less than 3%, it will cause a sharp increase in CO content, which greatly increases the chemical incomplete combustion heat loss. At the same time, it will also cause the increase of carbon content in fly ash, leading to the increase of mechanical incomplete combustion loss and the reduction of combustion efficiency. In addition, the low oxygen concentration can make some areas in the furnace become reducing atmosphere, thereby reducing the ash melting point caused furnace wall slagging and corrosion. Therefore, the use of low excess air combustion to reduce NO_x requires careful selection.

(2) Air-staged combustion

The traditional combustion method is to send all pulverized coal and air into the furnace and burn together. So pulverized coal and air fully mixed, high combustion intensity, high combustion temperature, but the resulting NO_x emissions are also very high. The air staged combustion technology is to control the mixing process of air and pulverized coal, the air required for combustion is fed into the combustion flame step by step, so as to realize the low oxygen combustion of pulverized coal particles in the early combustion, and achieve the purpose of reducing NO_x emissions. Air staged combustion technology includes vertical and horizontal air staged combustion.

The so-called vertical air staged combustion is to separate a part of the combustion air from the main burner and send it into the furnace from the upper part of the burner, which is called Over Fire Air (OFA). Due to the existence of OFA, the oxygen content in the main combustion zone decreases, the air volume decreases, and the formation of fuel NO_x decreases. Due to the decrease of combustion temperature, the formation of thermal NO_x decreases, so the total NO_x emission decreases.

Horizontal air staged combustion is to deflect the jet angle of the secondary wind and form a tangential circle of different sizes with the primary wind, delay the mixing of the secondary wind and the primary wind, and form a certain degree of air classification.

The use of air staged combustion technology for the transformation of old units is more convenient, the amount of change is small, and the cost of change is relatively low. After modification, the carbon content of fly ash increased and the boiler efficiency decreased. Since air staged combustion reduces the excess air coefficient in the main combustion zone, it is easy to increase the reducing atmosphere near the water wall, resulting in slagging and corrosion in the furnace. Therefore, this must be considered in the design to reduce the impact of this aspect. However, since air staged combustion reduces the combustion temperature level in the furnace, it is also beneficial to alleviate slagging in the furnace.

(3) Low NO_x burners (LNBs)

LNBs are designed to control air and fuel mixing to achieve staged combustion. Different from air staged combustion technology, LNB realizes staged air supply and fuel at burner outlet. The main advantage of LNBs is the reduction of NO_x generation as high as 30–50% and the easiness of their application in both existing and new furnaces. However, this procedure causes an increase of carbon content in ashes and greater CO formation [2]. Despite this, low NO_x burners are one of the most popular NO_x control technologies [3].

(4) In-furnace NO_x reduction (IFNR)

IFNR has been industrialized today. These burners provide stable flame with many zones, e. g. primary combustion zone, fuel reburning (FR) zone, and final combustion zone. The primary combustion zone is an oxidizing or weakly reducing atmosphere, in which the main fuel burns in an under-oxygen or weakly reducing environment to produce NO_x. In the fuel reburning zone, the secondary fuel is sent into the furnace to make it present a reducing atmosphere. Under high temperature and reducing atmosphere, the hydrocarbon atomic group is generated. The atomic group reacts with the NO_x generated in the primary combustion zone to reduce NO_x to N_2, which is usually called the reduction zone. The secondary fuel is usually called the reburning fuel. Above the reduction zone, a small amount of air is fed to make the reburning fuel burn completely, which is called final combustion zone, and the part of the secondary wind is called Over Fire Air.

2.1.2 *Post-combustion Methods*

Post-combustion methods are dealing with nitrogen oxides in exhaust gases from incineration processes. Currently, the most used post-combustion NO_x control methods are selective catalytic reduction (SCR) and selective non-catalytic reduction (SNCR), which are highlighted in the next.

(1) Selective noncatalytic reduction (SNCR)

Selective non-catalytic reduction is a method for NO_x reduction at 850–1100 °C without catalyst. Ammonia or urea is commonly used as the reducing agent in this technology. The reducing agent rapidly decomposes and reacts with NO_x in the flue gas to rapidly generate N_2 and H_2O. The main chemical reaction equations are

$$4NH_3 + 4NO + O_2 \rightarrow 4N_2 + 6H_2O \text{ (Ammonia as a reducing agent)}$$

$$(NH_2)_2CO \rightarrow 2NH_2 + CO \text{ (Urea as a reducing agent)}$$

$$NH_2 + NO \rightarrow N_2 + H_2O$$

$$2NO + 2CO \rightarrow N_2 + 2CO_2$$

When the temperature exceeds the reaction temperature window, ammonia will be oxidized to NO_x.

$$4NH_3 + 5O_2 \rightarrow 4NO + 6H_2O$$

This method appeared first in the mid1970s. In 1974 Wendt and Sterling injected NH_3 into the post flame gas and observed a reduction of NO [4, 5]. The SNCR is an attractive strategy to reduce NO_x emission due to its simplicity. Owing to a catalyst-free system all issues connected with the use of the catalyst present in SCR are absent in the SNCR. Thus, the capital and operation costs are lower. Further to the above, SNCR is easy to install in existing plants and can be applied in all types of stationary-fired equipment, however the main drawback of SNCR is its low efficiency from 30 to 75% [2].

(2) Selective catalytic reduction (SCR)

In the late 1970s, SCR technology was first applied to industrial boilers and power plant boilers in Japan, and Europe began to introduce SCR technology from 1985. The United States began to study SCR technology since 1959, and has obtained many patents in this area, but it did not develop to industrial applications until the late 1980s.

The mechanism of SCR technology is to convert NO_x into nitrogen and water by reducing agents (e.g. NH_3) at appropriate temperatures and in the presence of catalysts.

In the SCR process, the catalyst is placed in a box like a solid reactor. Catalyst units are usually arranged vertically, sometimes horizontally. Due to the mature technology and high denitrification rate, SCR technology has become the mainstream technology of flue gas denitrification in power plants in the world. With the increasing national environmental requirements, SCR technology has been widely used in China.

Detailed SCR method will be discussed in the following section.

2.2 SCR Process Configurations

The SCR catalysts in the form of honeycomb monoliths or plates are assembled into standard steel cased modules, which are then placed into the reactor to form catalyst layers (typically three layers), as shown in Fig. 2.1 [6]. Ammonia is vaporized, diluted with air, and injected into the flue gas stream through a distribution grid.

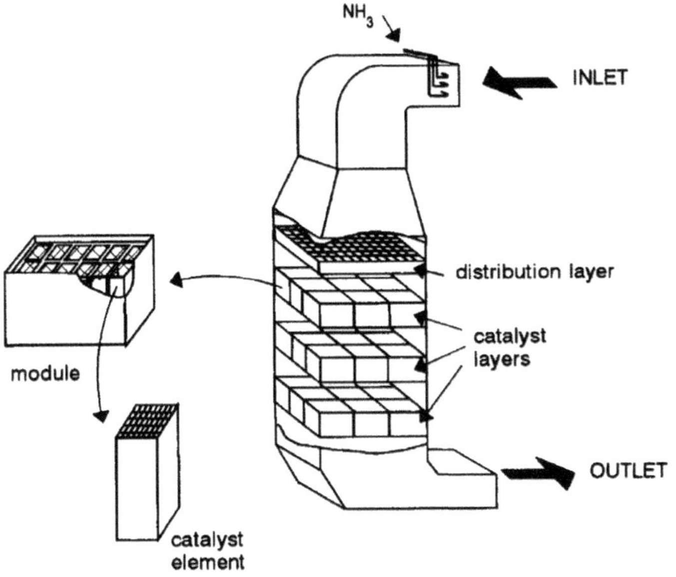

Fig. 2.1 The de-NO_x SCR reactor

The SCR reactor in thermal power plants can be located immediately after the boiler (high dust configuration, HD), after the removal of the particulate by the electrostatic precipitators (ESP) and upstream of the air pre-heater (low dust configuration, LD) and after the sulfur dioxide removal by a flue gas desulfurization (FGD) unit (tail end configuration, TE), as shown in Fig. 2.2.

2.2.1 HD-SCR Configuration

The HD-SCR configuration is the most common in coal-fired plants because the temperature of the flue gas between the economizer and the air pre-heater is optimal for catalyst activity (300–400 °C). But at this time all the fly ash and sulfur dioxide contained in the flue gas pass through the catalyst reactor, and the reactor works in the 'unclean' high dust gas. The life of the catalyst will be affected by the following factors.

(a) The fly ash carried by the flue gas contains Na, Ca, Si, As, and other components, which will poison or pollute the catalyst, thereby reducing the efficiency of the catalyst.
(b) The fly ash carried by the flue gas wears the SCR reactor.
(c) A large number of fly ash which contains in the flue gas is easy to block the honeycomb catalyst.

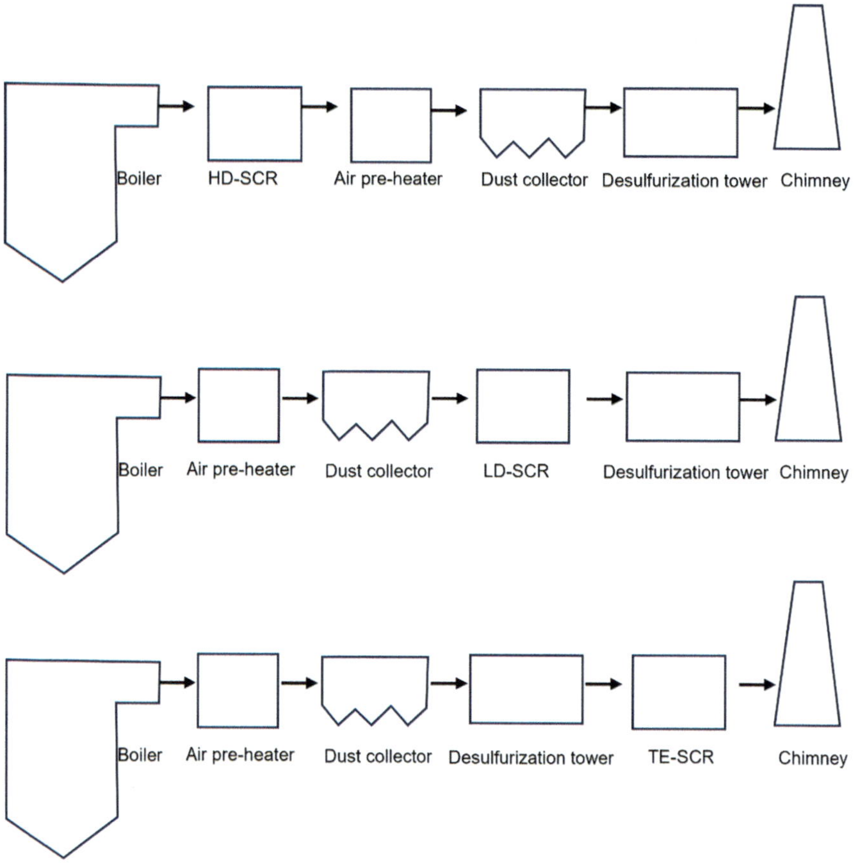

Fig. 2.2 Three configurations of SCR

(d) If the flue gas temperature rises, the catalyst will be sintered or recrystallization, resulting in catalyst deactivation. If the flue gas temperature decreases, NH_3 will react with SO_3 to form ammonium sulfate, thus blocking the SCR reactor channel and polluting the air pre-heater.
(e) Highly active catalysts promote the oxidation of SO_2 to SO_3 in flue gas.

Although there are many shortcomings, compared with other ways and considering other factors, HD-SCR configuration is still an economical and effective SCR arrangement.

2.2.2 LD-SCR Configuration

The SCR reactor is arranged behind the electrostatic precipitator. The flue gas first passes through the electrostatic precipitator and then enters the SCR reactor, which can prevent the fly ash in the flue gas from polluting the catalyst, wearing, or blocking the reactor, but SO_3 in the flue gas always exists. The biggest problem with LD-SCR configuration is that electrostatic precipitators cannot operate normally at 300–400 °C and are therefore rarely used.

2.2.3 TE-SCR Configuration

In the TE configuration the SCR reactor is arranged after the flue gas desulfurization (FGD), and the catalyst will work completely in the clean flue gas without dust and sulfur dioxide. Since there is no blockage and corrosion of the reactor by the fly ash, and there is no pollution and poisoning of the catalyst, the highly active catalyst can be used, which reduces the volume of the reactor and makes the reactor layout compact. When the catalyst works in clean flue gas, its working life can be up to twice the service life of the catalyst in HD-SCR reactor.

The main problem with this arrangement is that the reactor is arranged behind the wet FGD desulfurization unit, while the low-temperature SCR catalyst has not reached the degree of engineering application, and its exhaust gas temperature is below 150 °C. Therefore, in order to make the flue gas reach the required reaction temperature before entering the SCR reactor, extra heat is needed, thereby increasing energy consumption and operating costs.

Therefore, it is necessary to study the low-temperature SCR catalyst.

2.3 Low-Temperature SCR Catalyst

V_2O_5–$WO_3(MoO_3)/TiO_2$ is the efficient and typical catalyst and has been commercialized for NH_3-SCR technology for medium temperature process [7]. However, this catalyst has some intrinsic drawbacks such as narrow and high working temperature window (350–400 °C), and low N_2 selectivity in the high-temperature range [8–10]. Consequently, we should continue to develop highly active low-temperature NH_3-SCR catalysts, which can overcome most of the shortcomings of HD-SCR catalyst (typical catalyst). At the same time, a wide temperature window is also necessary.

Up to now, many low-temperature SCR catalysts have been investigated with excellent properties, low price, and high thermodynamic stability. For instance, Mn-based [11–13], Ce-based [13–16], Fe-based [17, 18], Cu-based [19], and Cr-based [18] catalysts were shown to express good SCR activity in low temperature. Nevertheless, these catalysts are sensitive to SO_2, H_2O, and metals (alkali metals, alkali-earth

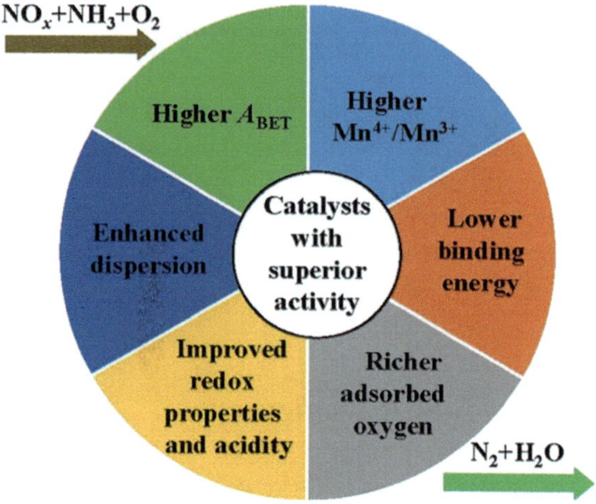

Fig. 2.3 General strategies for enhancing the activity of Mn-based catalysts. Reprinted with permission from Ref. [25]. Copyright 2016 Elsevier

metals, heavy metals) in fly ash [20–24]. Hence, there are many challenges in developing NH_3-SCR catalysts with high activity in low-temperature range, as well as resistance to SO_2, H_2O, and metals in fly ash.

Among these catalysts, Mn-based and Ce-based catalysts with superb performance have been extensively researched by many scholars.

Mn-based catalysts exhibit the most excellent low temperature SCR performance due to the following properties (as shown in Fig. 2.3):

(1) Mn-based catalysts have excellent redox properties and strong acidity. Remarkable oxidation–reduction ability is conducive to generating more active NH_2 and NO_2 species and optimizing the reaction pathway of NH_3-SCR. Strong acidity promotes the adsorption and activation of NH_3, which is a prerequisite for the formation of NH_2 species. Furthermore, strong acidity and outstanding oxidation–reduction ability are also two critical factors for the wide temperature window of SCR catalysts [26].
(2) Mn-based catalysts exhibit high N_2 selectivity at low temperatures, which could avoid secondary pollution caused by the generation of N_2O as much as possible.
(3) The optimized Mn-based catalysts have excellent SO_2/H_2O resistance at lower temperatures, which could inhibit the formation of sulfate or facilitate the decomposition of sulfate at low temperatures [27].
(4) Compared with noble metal catalysts, Mn-based catalysts are competitive in cost and easy availability. Therefore, they can be supplied in large quantities.

Ce-based catalysts have also been extensively researched as catalysts for low temperature NH_3-SCR, owing to the following properties:

(1) Cerium, exhibits two oxidation states of $+\ 3(Ce_2O_3)$ and $+\ 4(CeO_2)$ [28, 29]. The unique redox couple Ce^{4+}/Ce^{3+} can switch between CeO_2 and Ce_2O_3 under oxidizing and reducing conditions, respectively.

(2) Cerium oxide, which is most the rare-earth materials in the earth shell, is wide-source, non-toxic, and low-cost and can be supplied in large quantities.
(3) Ce-based catalysts perform excellent redox and large oxygen storage/release capacity. Moreover, labile oxygen vacancies and bulk oxygen species with relatively high mobility can be readily produced.
(4) It was corroborated that Ce was an effective additive to ameliorate low temperature NH_3-SCR activity.

In addition, Fe-based and Cu-based catalysts have also been extensively studied by researchers due to their unique properties in the fields of low temperature SCR catalyst.

2.4 Low-Temperature SCR Mechanism

During SCR reaction, NO_x can be converted into nitrogen and water naturally contained in the air by adding ammonia. The following is its main chemical equation:

$$4NO + 4NH_3 + O_2 \rightarrow 4N_2 + 6H_2O$$

$$6NO + 4NH_3 \rightarrow 5N_2 + 6H_2O$$

$$6NO_2 + 8NH_3 \rightarrow 7N_2 + 12H_2O$$

$$2NO_2 + 4NH_3 + O_2 \rightarrow 3N_2 + 6H_2O$$

The above equations are only the overall reaction result, and the specific reaction mechanism will be explained from the following part.

As is well known that the adsorption behaviors of gaseous on the surface of catalyst are the important processes in the gas–solid phase catalytic reaction [30]. Therefore, the gaseous molecules ($NO + O_2$, NO_2 and NH_3) are adsorbed on the surface of low temperature SCR catalysts, which plays an important role in low temperature SCR mechanism. Numerous researchers have studied low temperature NH_3-SCR mechanism and they believed that the reaction pathways of low temperature NH_3-SCR catalysts could be divided into two types (E–R and L–H mechanisms) [23, 27, 31–39].

2.4.1 Eley–Rideal (E–R) Mechanism

NH_3 was absorbed on Lewis acid sites and oxidized to NH_2, and then the formed NH_2 reacted with gaseous NO generating N_2 and H_2O.

$NH_3(g) \rightarrow NH(a)$ (Lewis acid sites)
$O_2(g) \rightarrow 2O(a)$
$NH_3(a) + O(a) \rightarrow NH_2(a) + OH(a)$
$NO(g) + NH_2(a) \rightarrow NH_2NO(a)$
$NH_2NO(a) \rightarrow N_2(g) + H_2O$.

2.4.2 *L–H Mechanism*

L–H mechanism: the adsorbed NO_2 could react with coordinated NH_3 in the NH_3-SCR reaction.

$NO + O_2(g) \rightarrow NO_2(a)$
$NH_3(g) \rightarrow NH_3(a)$ (Lewis acid sites)
$NO_2(a) + 2NH_3(a) + NO(g) \rightarrow 2N_2 + 3H_2O$.

Numerous researchers have systematically studied the adsorption and interaction of NO_x and NH_3 about the Ce-based and Mn-based catalysts.

Yang et al. proposed that the process of the NH_3-SCR reaction of pure CeO_2 followed the E–R mechanism [40]. Subsequently, Sun et al. contrasted the contributions of the E–R mechanism and the L–H mechanism over the Ce/TiO_2 catalyst [41]. The experimental results revealed that the NH_3-SCR reaction mechanism of the Ce/TiO_2 catalyst was chiefly regulated by the L–H mechanism at a low temperature (below 100 °C). Furthermore, the proportion of the E–R mechanism heightened rapidly with increasing temperature (Fig. 2.4).

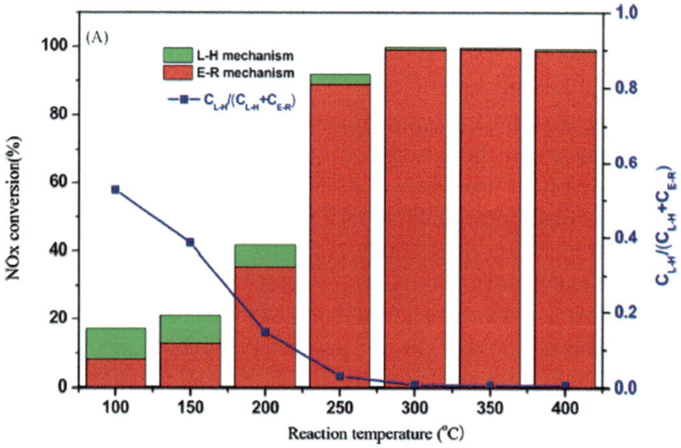

Fig. 2.4 Contributions of the E–R and L–H mechanisms to the NH_3-SCR reaction over Ce/TiO_2 at different temperatures. Reaction conditions: [NO] = [NH_3] = 600 ppm, [O_2] = 5%, GHSV = 108,000 h^{-1}. Reprinted from Ref [41]. Copyright 2017, with permission from Elsevier

A large number of studies and chemical calculations showed that the reaction mechanisms of different kinds of MnO_x-based catalysts were quite different. This was mainly due to the different redox performance and surface acidity of Mn-based catalysts in different systems, which could produce different NH_x/NO_x active intermediates [42]. Yang et al. [43] synthesized a Mn–Ce composite catalyst by a hydrothermal method. According to the analysis of in situ DRIFTS, when NO + O_2 was imported to react with pre-adsorbed NH_3, the bands corresponding to NH_4^+ (1428 cm^{-1}) and $-NH_2$ (1547 cm^{-1}) gradually disappeared with increasing exposure time. In contrast, the bands corresponding to monodentate nitrite (1372 cm^{-1}) gradually increased. This confirmed that NH_4^+ and $-NH_2$ reacted with gaseous NO(g) rather than adsorbed nitrate and other species, and the NH_3-SCR reaction followed the E–R mechanism [44].

From above discussion, in most cases, a catalyst does not follow only one reaction mechanism, and a combination of L–H and E–R mechanisms influence the low temperature SCR performance of the catalyst.

2.5 Application of Density Functional Theory (DFT) in the Study of Low Temperature SCR Catalyst

Over recent years, the application of quantum chemical calculation using density functional theory (DFT) has been developed rapidly in the field of SCR for the mechanisms of reaction and poisoning. A typical application of DFT in the study of low temperature SCR catalyst is visualization of reaction and poisoning mechanisms and design of functional materials.

The SCR reaction mechanism over Mn–CeO_2 catalysts has been investigated by DFT calculations [45, 46]. By DFT + U calculation, Song et al. [46] found that NH_3 was adsorbed on the Lewis acid Mn sites and further dissociated one of its N–H bonds to form the key NH_2 intermediate. NO adsorption on this NH_2 intermediate resulted in the form of NH_2NO that could then undergo further N–H cleavage reactions to form OH groups. They proposed that the redox mechanism involved doped Mn, as Lewis acid sites for ammonia adsorption and O vacancies in the ceria surface makes N_2O decompose into the desired N_2 product. Peng et al. [47, 48] researched As_2O_5 and alkali metals poisoning resistant capacity of V_2O_5–WO_3–TiO_2 by using DFT theoretical approaches. They found that the decrease of acidity after poisoned might directly reduce the catalytic activity calculated by the DOS and PDOS, and the alkali atom mainly influences the active site V species rather than W species [47]. Phil et al. [49] inferred that the catalyst of high resistance to SO_2 poisoning should be combined with sulfate weakly using quantum chemical calculation, which means that the bonding strength of metal atom and oxygen atom is weaker. In this theory, they selected the transition metal Sb and prepared Sb–V_2O_5/TiO_2 catalyst of high resistance to H_2O and SO_2 poisoning. Maitarad et al. [45] constructed the

Mn–CeO_2 (110) model on the basis of a combination of experiment and calculation. DFT calculations demonstrated that the catalyst has excellent catalytic activity and NO and NH_3 can be adsorbed on the catalyst surface. Lu et al. [50] proposed that the formation of bidentate sulfates was mainly generated on the plane (110) of the CeO_2 surface during the SCR process by means of density of states, electron localization functions and charge density difference. Liu et al. [51] investigated the SO_2 poisoning mechanism of MnxC$e_{1-x}O_2$ catalyst used CeO_2 as the model by the method of calculation. By calculating the electronic structure, density of states and the energy barrier that SO_2 needs to overcome in the formation of sulfates of CeO_2 and $Mn_xCe_{1-x}O_2$ being obtained, it was noted that the CeO_2 deactivation was due to the formation of $CeSO_4$ rather than $Ce_2(SO_4)_3$, which was acidized by SO_2. For the $Mn_xCe_{1-x}O_2$ catalyst, it was easier to be poisoned since the introduction of Mn enhanced the thermal stability of the surface sulfate.

Guo et al. [20, 21] also studied that the alkali metals and heavy metals poisoning Mn/TiO_2 catalyst by DFT calculation. They found that the alkali and heavy metals changed the surface of the catalysts and the adsorption energy of NH_3 and NO become smaller than pure Mn/TiO_2 catalyst.

As we know that DFT calculation will be more widely used in low temperature SCR in the future.

References

1. Friebel J, Köpsel RFW. The fate of nitrogen during pyrolysis of German low rank coals—a parameter study. Fuel. 1999;78:923–32.
2. Gomez-Garcia MA, Pitchon V, Kiennemann A. Pollution by nitrogen oxides: an approach to NO_x abatement by using sorbing catalytic materials. Environ Int. 2005;31:445–67.
3. Ballester JM, Dopazo C, Fueyo N, et al. Investigation of low-NO_x strategies for natural gas combustion. Fuel. 1997;76(5):435–46.
4. Javed TM, Irfan N, Gibbs BM. Control of combustion-generated nitrogen oxides by selective non-catalytic reduction (review). J Environ Manage. 2007;83:251–89.
5. Muzio LJ, Quartucy GC. Implementing NO_x control: research to application. Prog Energ Combust. 1997;23:233–66.
6. Forzatti P. Present status and perspectives in de-NO_x SCR catalysis. Appl Catal A. 2001;222:221–36.
7. Boningari T, Koirala R, Smirniotis PG. Low-temperature selective catalytic reduction of NO with NH_3 over V/ZrO_2 prepared by flame-assisted spray pyrolysis: structural and catalytic properties. Appl Catal B. 2012;127:255–64.
8. Xin Y, Li Q, Zhang Z. Zeolitic materials for DeNO_x selective catalytic reduction. ChemCatChem. 2018;10:29–41.
9. Chen L, Li J, Ge M. DRIFT study on cerium-tungsten/titiania catalyst for selective catalytic reduction of NO_x with NH_3. Environ Sci Technol. 2010;44:9590–6.
10. Liu F, Asakura K, He H, et al. Influence of sulfation on iron titanate catalyst for the selective catalytic reduction of NO_x with NH_3. Appl Catal B. 2011;103:369–77.
11. Li Y, Guo R, Wang Z, et al. Removal of NO by using sodium persulfate solution: catalyzed by different nanostructured MnO_2. Energy Fuels. 2020;34:3477–82.
12. Kleinm L, Lee YN, Springston SR, et al. Ozone formation at a rural site in the southwestern United States. J Geophys Res. 1994;99:3469–82.

13. Yao X, Kong T, Chen L, et al. Enhanced low-temperature NH_3-SCR performance of MnO_x/CeO_2 catalysts by optimal solvent effect. Appl Surf Sci. 2017;420:407–15.
14. Duan C, Guo R, Liu Y, et al. Enhancement of potassium resistance of Ce–Ti oxide catalyst for NH_3-SCR reaction by modification with holmium. J Rare Earths. 2020;40(1):49–56.
15. Duan C, Guo R, Wu G, et al. Selective catalytic reduction of NO_x by NH_3 over $CeVO_4$-CeO_2 nanocomposite. Environ Sci Pollut Res Int. 2020;27(18):22818–28.
16. Guo R, Chen Q, Ding H, et al. Preparation and characterization of CeO_x@MnO_x core–shell structure catalyst for catalytic oxidation of NO. Catal Commun. 2015;69:165–9.
17. Huang C, Guo R, Pan W, et al. SCR of NO_x by NH_3 over $MnFeO_x$@TiO_2 catalyst with a core-shell structure: the improved K resistance. J Energy Inst. 2019;92(5):1364–78.
18. Sun X, Guo R, Li M, et al. The promotion effect of Fe on $CeZr_2O_x$ catalyst for the low-temperature SCR of NO_x by NH_3. Res Chem Intermed. 2018;44(5):3455–74.
19. Pan W, Hong J, Guo R, et al. Effect of support on the performance of Mn–Cu oxides for low temperature selective catalytic reduction of NO with NH_3. J Ind Eng Chem. 2014;20(4):2224–7.
20. Guo R, Wang Q, Pan W, et al. The poisoning effect of heavy metals doping on Mn/TiO_2 catalyst for selective catalytic reduction of NO with NH_3. J Mol Catal A: Chem. 2015;407:1–7.
21. Guo R, Wang Q, Pan W, et al. The poisoning effect of Na and K on Mn/TiO_2 catalyst for selective catalytic reduction of NO with NH_3: a comparative study. Appl Surf Sci. 2014;317:111–6.
22. Guo R, Wang S, Pan W, et al. Different poisoning effects of K and Mg on the Mn/TiO_2 catalyst for selective catalytic reduction of NO_x with NH_3: a mechanistic study. J Phys Chem C. 2017;121(14):7881–91.
23. Liu J, Guo R, Li M, et al. Enhancement of the SO_2 resistance of Mn/TiO_2 SCR catalyst by Eu modification: a mechanism study. Fuel. 2018;223:385–93.
24. Wang S, Guo R, Pan W, et al. The deactivation mechanism of Pb on the Ce/TiO_2 catalyst for the selective catalytic reduction of NO_x with NH_3: TPD and DRIFT studies. Phys Chem Chem Phys. 2017;19(7):5333–42.
25. Wang X, Wu S, Zou W, et al. Fe-Mn/Al_2O_3 catalysts for low temperature selective catalytic reduction of NO with NH_3. Chin J Catal. 2016;37:1314–23.
26. Yu Y, Tan W, An D, et al. Insight into the SO_2 resistance mechanism on γ-Fe_2O_3 catalyst in NH_3-SCR reaction: a collaborated experimental and DFT study. Appl Catal B. 2021;281:119544.
27. Zhang Z, Li R, Wang M, et al. Two steps synthesis of $CeTiO_x$ oxides nanotube catalyst: Enhanced activity, resistance of SO_2 and H_2O for low temperature NH_3-SCR of NO_x. Appl Catal B. 2021;282:119542.
28. Gschneidner KA, Maple MB, Eyring LR. Handbook on the physics and chemistry of rare earths. Elsevier Sci Technol. 2000;57:295–300.
29. Liu S, Wu X, Weng D, et al. Ceria-based catalysts for soot oxidation: a review. J Rare Earth. 2015;33:567–90.
30. Kijlstra WS, Brands DS, Poels EK, et al. Mechanism of the selective catalytic reduction of NO by NH_3 over MnO_x/Al_2O_3. J Catal. 1997;171:208–18.
31. Zhang L, Pierce J, Leung V, et al. Characterization of Ceria's interaction with NO_x and NH_3. J Phys Chem C. 2013;117:8282–9.
32. Chen Y, Wang J, Yan Z, et al. Promoting effect of Nd on the reduction of NO with NH_3 over CeO_2 supported by activated semi-coke: an in situ DRIFTS study. Catal Sci Technol. 2015;5:2251–9.
33. Li W, Guo R, Wang S, et al. The enhanced Zn resistance of Mn/TiO_2 catalyst for NH_3-SCR reaction by the modification with Nb. Fuel Process Technol. 2016;154:235–42.
34. Wang S, Guo R, Pan W, et al. The deactivation mechanism of Pb on the Ce/TiO_2 catalyst for the selective catalytic reduction of NO_x with NH_3: TPD and DRIFT studies. Phys Chem Chem Phys. 2017;19:5333–42.
35. Wang S, Guo R, Pan W, et al. The deactivation of Ce/TiO_2 catalyst for NH_3-SCR reaction by alkali metals: TPD and DRIFT studies. Catal Commun. 2017;89:143–7.
36. Guo R, Li M, Sun P, et al. Mechanistic investigation of the promotion effect of Bi modification on the NH_3-SCR performance of Ce/TiO_2 catalyst. J Phys Chem C. 2017;121:27535–45.

37. Wang Z, Guo R, Guan Z, et al. The promotion effect of Cr additive on $CeZr_2Ox$ catalyst for the low temperature selective catalytic reduction of NO_x with NH_3. Appl Surf Sci. 2019;485:133–40.
38. Yang N, Guo R, Pan W, et al. The deactivation mechanism of Cl on Ce/TiO_2 catalyst for selective catalytic reduction of NO with NH_3. Appl Surf Sci. 2016;378:513–8.
39. Guo R, Sun X, Liu J, et al. Enhancement of the NH_3-SCR catalytic activity of $MnTiO_x$ catalyst by the introduction of Sb. Appl Catal A. 2018;558:1–8.
40. Yang S, Guo Y, Chang H, et al. Novel effect of SO_2 on the SCR reaction over CeO_2: mechanism and significance. Appl Catal B. 2013;136−137:19−28
41. Sun P, Guo R, Liu S, et al. Enhancement of the low-temperature activity of Ce/TiO_2 catalyst by Sn modification for selective catalytic reduction of NO_x with NH_3. Mol Catal. 2017;433:224–34.
42. Li G, Wang B, Wang Z, et al. Reaction mechanism of low-temperature selective catalytic reduction of NO_x over Fe-Mn oxides supported on fly-ash-derived SBA-15 molecular sieves: structure-activity relationships and in situ DRIFT analysis. J Phys Chem C. 2018;122:20210–31.
43. Yang C, Yang J, Jiao Q, et al. Promotion effect and mechanism of MnO_x doped CeO_2 nano-catalyst for NH_3-SCR. Ceram Int. 2020;46:4394–401.
44. Xiao X, Wang J, Jia X, et al. Low-temperature selective catalytic reduction of NO_x with NH_3 over Mn–Ce composites synthesized by polymer-assisted deposition. ACS Omega. 2021;6:12801–12.
45. Maitarad P, Zhang D, Gao R, et al. Combination of experimental and theoretical investigations of $MnO_x/Ce_{0.9}Zr_{0.1}O_2$ nanorods for selective catalytic reduction of NO with ammonia. J Phys Chem C. 2013;117:9999–10006.
46. Song W, Liu J, Zheng H, et al. A mechanistic DFT study of low temperature SCR of NO with NH_3 on $MnCe_{1-x}O_2(111)$. Catal Sci Technol. 2016;6:2120–8.
47. Peng Y, Li J, Shi W, et al. Design strategies for development of SCR catalyst: improvement of alkali poisoning resistance and novel regeneration method. Environ Sci Technol. 2012;46:12623–9.
48. Peng Y, Li J, Si W, et al. Insight into deactivation of commercial SCR catalyst by arsenic: an experiment and DFT study. Environ Sci Technol. 2014;48:13895–900.
49. Phil H, Reddy M, Kumar P, et al. SO_2 resistant antimony promoted V_2O_5/TiO_2 catalyst for NH_3-SCR of NO_x at low temperatures. Appl Catal B. 2008;78:301–8.
50. Lu Z, Müller C, Yang Z, et al. SO_x on ceria from adsorbed SO_2. J Chem Phys. 2011;134:184703–14.
51. Liu Y, Cen W, Wu Z, et al. SO_2 poisoning structures and the effects on pure and Mn doped CeO_2: a first principles investigation. J Phys Chem C. 2012;116:22930–7.

Chapter 3
Preparation of Catalysts

3.1 General Process of NH_3-SCR Catalysts Preparation

The two primary components of the NH_3-SCR catalytic process are the preparation and use of industrial catalysts, and only through proper preparation and use of high-quality catalysts can we utilize their full potential for an effective industrial catalytic process [1]. The conventional catalyst preparation process is shown in Fig. 3.1.

It is challenging to achieve the requirements for industrial NH_3-SCR catalysts, which include high activity, good selectivity, stability under service circumstances, strong thermal stability, mechanical stability, anti-toxic qualities, and low cost. The catalysts can be prepared in a variety of methods. Even though the raw ingredients and dosage are the same, the performance of the catalysts produced can vary significantly depending on the procedure. Many microscopic elements in the industrial catalyst preparation process are challenging to manage because of their intricacy. Although the requirements for the prepared catalysts are obviously gradually improving, the difficulty of catalyst preparation has now been substantially reduced as a result of the accumulation of theoretical research and actual production experience [2]. At present, the main methods used for the preparation of low-temperature NH_3-SCR catalysts are precipitation, sol–gel, impregnation, and hydrothermal methods. This is followed by separate presentations:

3.2 Precipitation Method

Precipitation is the precipitation reaction that transforms soluble catalyst components (aqueous solutions of metal salts) into challenging compounds (e.g. alkaline substances). Precursor salt selection, preparing a solution with the correct solute concentration, precipitation, and ageing at the desired pH, temperature, and pressure are the typical steps. After that, several procedures may be repeated in additional

W. Pan and R. Guo, *Low-Temperature Selective Catalytic Reduction Catalysts*,
https://doi.org/10.1007/978-981-97-9298-6_3

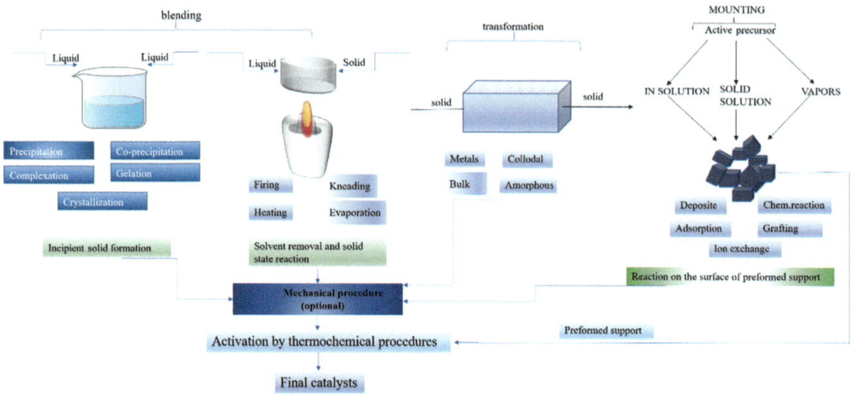

Fig. 3.1 Simplified diagram summarizing methods of blending, transforming, and mounting catalyst components

preparation techniques, such as washing, drying, comminution, shaping, and thermal processing such as calcination and activation.

3.2.1 *Precipitating Classical Theory*

The chemical process of precipitation is intricate. When a precipitant interacts with an aqueous solution of a metal salt, a precipitate results when the concentration of the precipitating ion is higher than the corresponding condition's solubility product. It is crucial to comprehend the fundamentals of sedimentation theory to produce a well-structured and pure precipitate. The formation of precipitates includes two processes: the generation and the growth of nuclei. The former process is the formation of precipitate ions that collide with each other to generate the nuclei of precipitation, nuclei in aqueous solution in the equilibrium of precipitation and dissolution, the specific surface area is large, and thus the solubility of the precipitate than the grain is large, the formation of a supersaturated solution, if the saturated concentration of solute at a certain temperature is $\mathbf{c^*}$, the concentration in the over billion and solution is $\mathbf{c}$, then $\mathbf{S = c/c^*}$ is called the degree of supersaturation [3]. When the solution reaches a specific level of supersaturation and the rate of solid phase production is greater than the rate of solid phase dissolution, a huge number of nuclei are formed instantly. This is when crystals begin to form. The solution's solute molecules then diffuse to the nuclei's surface as they proceed to form crystals (Fig. 3.2) [4, 5].

(1) **Thermodynamics of nucleation**

In the nucleation process, two particles of a given substance (atom, molecule, ion). To produce a new phase, a critical nucleus must be formed [6]. Only then can the critical nuclei grow spontaneously into macroscopic crystalline size. The formation of

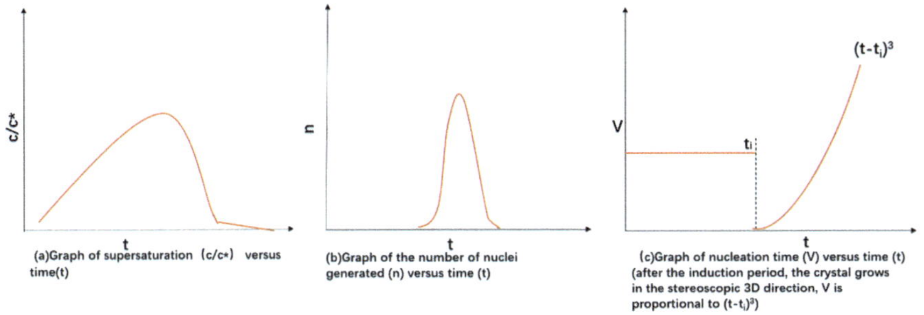

Fig. 3.2 Rate of formation of insoluble precipitates. nucleation starts at ti, the induction time, where the largest number of nuclei is formed instantaneously, followed by a rapid decrease in the number of newly formed nuclei

a critical nucleus requires work, which is equivalent to the Gibbs free energy change of the system. It comprises of the formation energy of the new phase interface and two chemical potential differences between the two states. Due to the discontinuity in the concentrations of solution components at the interface, the development of such an interface requires a free energy cost. The lower the free energy, the faster the formation of the condensed phase (Fig. 3.3) [7–9].

(2) **Kinetics of nucleation**

Nucleation is the creation of a new phase. Depending on the conditions of the system, the process may proceed according to any of the following processes [12] (Fig. 3.4).

The primary nucleation process is characterized by the production of a solid phase without precipitation crystallization, while secondary nucleation is the formation of new particles in the presence of their crystallization [13]. Secondary nucleation is the process of forming new particles in the presence of crystallization of their matter.

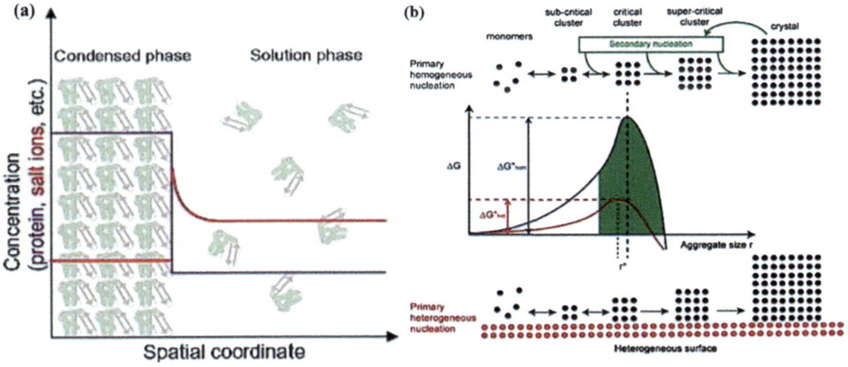

Fig. 3.3 **a** The nucleation of a new phase from Ref. [10] and **b** the creation of crystals from solution follows either through primary (homogeneous or heterogeneous) or through secondary nucleation from Ref. [11]

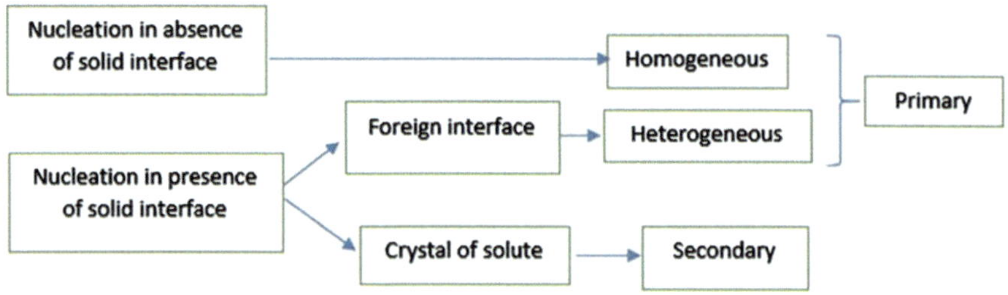

Fig. 3.4 Schematic diagram of possible nucleation processes from Ref. [12]

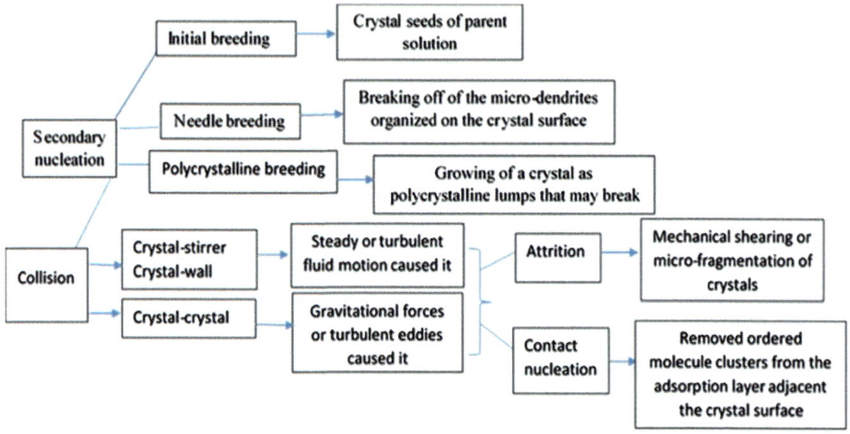

Fig. 3.5 Classification of secondary nucleation mechanisms

The process is the formation of nuclei in a medium that does not contain a solid phase. Multiphase nucleation is due to the presence of and takes place at the phase boundaries within the system (colloids etc.). The nucleation takes place on the phase boundary of the system (colloids etc.). For the precipitation of refractory solid phases from aqueous solutions, the majority are fine crystals. The mechanical breakage is unlikely, so it is probably not as easy to achieve secondary nucleation as with easy-bathing substances. Secondary nucleation is not as easy as with easy-bath materials. In the case of zeolite molecular sieve synthesis systems, these nucleation processes often coexist (Fig. 3.5). So, in practice, these processes are intertwined [11, 14].

3.2.2 Factors Affecting Catalyst Performance in the Precipitation Method

(1) **Effect of concentration**

It has already been mentioned that the pace of nucleation and nucleation growth during the precipitation formation process determines the shape of the precipitate that is produced. In turn, concentration depends on how much the rate of nucleation is greater than the rate of nuclei growth. They fall into the following categories:

Nucleation rate: Nucleation is the process of generating a new phase, which can only occur when solute molecules or ions have sufficient energy to overcome the resistance at the liquid–solid interface. The nuclei can only be formed when the solute molecules or ions have sufficient energy to overcome the resistance at the liquid–solid interface.

Core growth rate: The nucleation process is similar to other mass transfer processes with chemical reactions and can be divided into two steps: firstly, the solute molecules diffuse through the stagnant layer at the liquid–solid interface; secondly, a surface reaction takes place and the molecules or ions are accepted into the lattice.

(2) **Effect of temperature**

It was previously proposed that the production and expansion of nuclei, which are in turn tightly correlated with temperature, are directly influenced by the supersaturation of a solution. Increased temperature reduces the degree of supersaturation and hence lowers the rate of nucleation when the amount of solute in a solution is constant; conversely, decreased temperature raises the degree of supersaturation and consequently raises the rate of nucleation. However, when the energy impact is taken into account, the relationship between these variables will become more complicated. As the temperature the energy of the solute molecules is very low at low temperatures, the rate of nucleation remains small. As the temperature increases, the rate of nucleation can reach very high values as motioned in Fig. 3.2a.

(3) **Effect of pH of the solution**

In precipitation processes, alkaline compounds are frequently utilized as precipitating agents. Changes in the solution's pH, of course, have an impact on the precipitation process [15]. For example, if aluminium salts are precipitated by alkali, three products can be obtained when other conditions are the same and the pH varies (Fig. 3.6) [16].

(4) **Effect of precipitating agents**

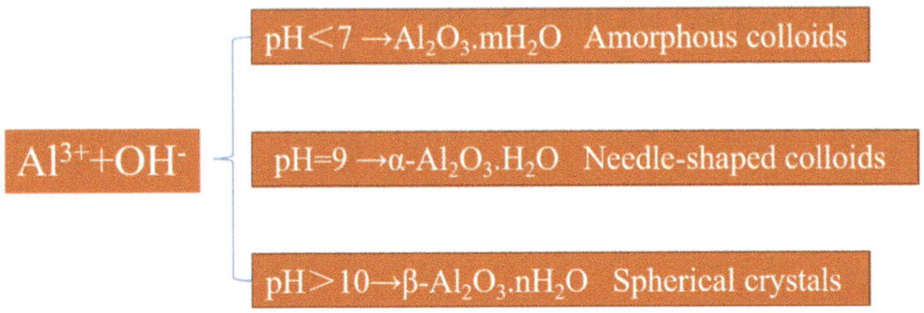

Fig. 3.6 Three aluminium salts are obtained when other conditions are the same and the pH varies

Whenever possible, use substances that are easily decomposed and contain volatile components as precipitating agents. Commonly used precipitants include bases (sodium hydroxide, potassium hydroxide, etc.), urea, ammonia, press salts (carbonate, bicarbonate, sulphate, oxalate, etc.), carbon dioxide, carbonates (sodium carbonate, potassium carbonate, sodium bicarbonate), etc. Of these, sodium carbonate and sodium hydroxide are the more common precipitants.

3.2.3 Brief Conclusion

A mixture of solutions containing two or more metal ions interacting with a precipitant to form a precipitate containing several metal components at the same time is known as the precipitation method. The use of precipitation allows the preparation of multi-component NH_3-SCR catalysts, which is one of the most common methods. The principle of precipitation is essentially the same as that of single-component precipitation, but the composition of the precipitates is more complex and the different precipitation conditions can result in significantly uneven precipitation products due to the different solubility products of the components. Therefore, although the precipitation method is a simple process, the preparation conditions and influencing factors of the corresponding NH_3-SCR catalyst must be carefully considered when using this method.

3.3 Sol–Gel Method

Sol–gel method is employed when a material's textural qualities must be highly controlled and the components must be disseminated on nearly molecular size. For the NH_3-SCR catalysts, it is an easy way to prepare materials with a high degree of dispersion. In contrast to more traditional industrial methods, the sol–gel process is typically referred to as "soft chemistry" [17]. Nowadays, the term 'sol–gel' is used more broadly to cover the synthesis of solid materials, such as metal oxides, from solution-state precursors. The development of technologically advanced synthesis routes that overcome the obstacles that are encountered in the most traditional procedures allowed the sol–gel technique to become a unique tool in the design of metal oxides with controlled architecture [18, 19].

3.3.1 Fundamentals of Sol–Gel Process

High-purity compounds can be created by using precursors with low impurity concentrations. A sol–gel is a colloidal dispersion of solid particles in a liquid. In a suspension known as a colloid, the dispersed phase is so minute (~ 1 nm–1 μm) that gravitational forces are negligible and interactions are mostly governed by short-range effects [20]. Gels can be regarded as consisting of a continuous colloidal-sized solid and liquid phase. A line segment that leaves a pore and penetrates the closest solid surface vertically must reappear in a poreless than 1 μm distant since both phases are colloidal. Similarly, a line segment that exits the solid phase and enters the pore wall vertically must return to the solid phase 1 μm later [21–23].

Producing supported metal catalysts with the necessary physicochemical characteristics that result in the features needed for practical applications is crucial. For a material to have the desired properties, such as high activity, high selectivity, chemical, thermal, and mechanical stability, rapid mass and heat transfers, high activity, high dispersion, adequate composition when formed into an alloy, low mobility, and support, including high and well-distributed porosity, are all required [24]. Depending on whether the liquid in the wet gel is removed by evaporative drying or by supercritical drying, that is in pressure and temperature conditions beyond the critical point of the liquid, the resulting dry material is named xerogel or aerogel respectively. Figure 3.7 shows the schematic diagram of the sol–gel method which can be summarized, as follows:

- Preparation of the solution of precursors.
- Hydrolysis and partial condensation of alkoxides to form a "sol".
- Formation of the gel via polycondensation of hydrolyzed precursors.
- Drying. The gel forms a dense "xerogel" via the collapse of the porous network caused by the evaporation of the solvent (or an aerogel for example through supercritical drying).
- Calcination to obtain mechanically stable materials.

An ageing stage may be included since chemical interactions carry on long after the gel is formed, changing the content, structure, and characteristics of the gel. The various steps can be lengthened, changed, or adjusted depending on the particular application to meet various needs. Sol–gel technique variables include (1) precursor concentration and type; (2) solvent type; (3) pH of the solution; (4) additive type and concentration (catalysts, surfactants, structure directing agents); (5) pre- and post-heating of the components; and (6) ageing period [25].

3.3.2 Sol–Gel Methods for Preparing Supported Metals

The common industrially important preparation methods of supported metal catalysts are multi-processes consisting of [23]

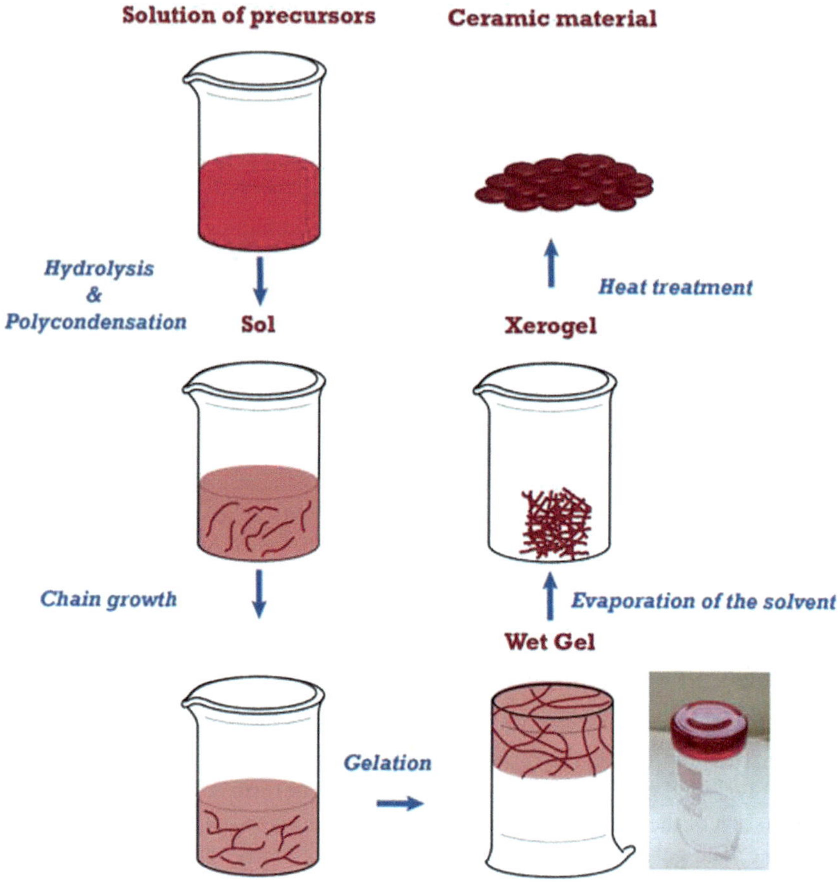

Fig. 3.7 Schematic representation of the sol–gel process

(a) preparation of the support;
(b) distribution of the active component precursor over the support surface,
(c) drying, calcination or pyrolysis and possibly reduction of the catalyst to obtain the active species (often metal or metal oxide nanoparticles) dispersed in the porosity of the support.

The ability to create a solid from a homogenous solution that comprises both the support precursor(s) and the metal precursor(s) is a significant benefit of the sol–gel method. The steps (a) and (b) associated with traditional approaches are combined into a single step. As was already indicated, one benefit of making supported metals using the sol–gel technique is that the metal can be added before the development of the support. In addition to saving a step and being more convenient, this characteristic enables the preparation of supported metal catalysts with distinctive structures that are not possible with other preparation techniques. When compared to supported catalysts made via impregnation, one-step sol–gel techniques frequently result in larger dispersion [26, 27].

3.3.3 *Brief Conclusion*

Although the sol–gel method is particularly attractive for the preparation of low-temperature NH_3-SCR catalysts because it allows for the preparation of the carrier and active sites in a single step, the two-step method, which involves the dispersion of the active metal precursor in the pores of an already prepared hetero-gel or aerogel carrier, has been shown to produce catalysts with better performance than catalysts prepared in the one-step method under certain specific circumstances. The one-step technique, although it often results in the preparation of widely dispersed supported metal catalysts, occasionally leads to poor dispersion of metal species [28]. In addition, most of the carriers of one-step catalysts may suffer from occlusion of the active metal, which will result in partial or complete inaccessibility of the reactants to the carriers during the catalytic process, depending on the preparation and/or operating conditions of the catalysts, such as temperature, pressure, or composition of the gas environment. Therefore, the use of the sol–gel method in a two-step process for the production of effective NH_3-SCR catalysts is sometimes preferable. In conclusion, the sol–gel method opens up new avenues for the synthesis of NH_3-SCR catalysts due to its highly controllable nature.

3.4 Impregnation Method

Impregnation is the process of submerging the carrier in a solution that contains the active ingredient. This is a typical technique for creating multi-component NH_3-SCR catalysts and entails three procedures: (1) The conception procedure. Under specific circumstances, an impregnating solution containing the active component is dissolved and applied to the dry or wet carrier to impregnate it. (2) The drying procedure. At a specific temperature, the solvent separates from the impregnated catalyst. (3) The activation procedure. By roasting the catalyst with air at a specific temperature or by employing a reducing agent like hydrogen, the NH_3-SCR catalyst is activated [29]. In addition, wet impregnation uses an excessive amount of solution, which is separated from the solid by drying over a predetermined time. The solution's composition is changed during the diffusion process, resulting in the formation of an impurity mud and the rapid release of heat from adsorption. The metallic catalyst can be effectively prepared using the wet-impregnation approach to achieve high performance and a stable microstructure.

3.4.1 Impregnating Solution Preparation

In Fig. 3.8 are shown four limiting types of impregnation profiles. Proper control of the operating conditions during impregnation and drying determines the impregnation profile of the final catalyst. Solution impregnation usually an aqueous solution of soluble salts of the required active component is prepared. Vapour phase impregnation is usually carried out using metal salts or oxides of the active component ($AlCl_3$, B_2O_3, MoO_3, etc.) which have a low boiling point and sublimate at low temperatures to precipitate the active component on the carrier in the vapor phase, as in the case of industrial n-butyl isomerization catalysts. Depending on the impregnation method, these can be divided into supersaturated and saturated impregnations. Depending on the carrier, dry impregnation and wet impregnation can be used. Dry impregnation is carried out with a dry carrier, while wet impregnation is carried out with a carrier that is pre-impregnated with water and then impregnated with a solution of active components. The impregnation method has several characteristics: the process is simple, especially the granular carrier impregnation method has a large processing capacity and a high production capacity; And the high-temperature treatment of the carrier is particularly beneficial for improving the activity and stability of the catalyst.

When a carrier with a porous structure is impregnated with a solution containing active components, the solution is drawn from the surface into the fine pores of the carrier by capillary forces, and the active components of the solute penetrate and diffuse towards the inner walls of the fine pores and are then adsorbed by the active

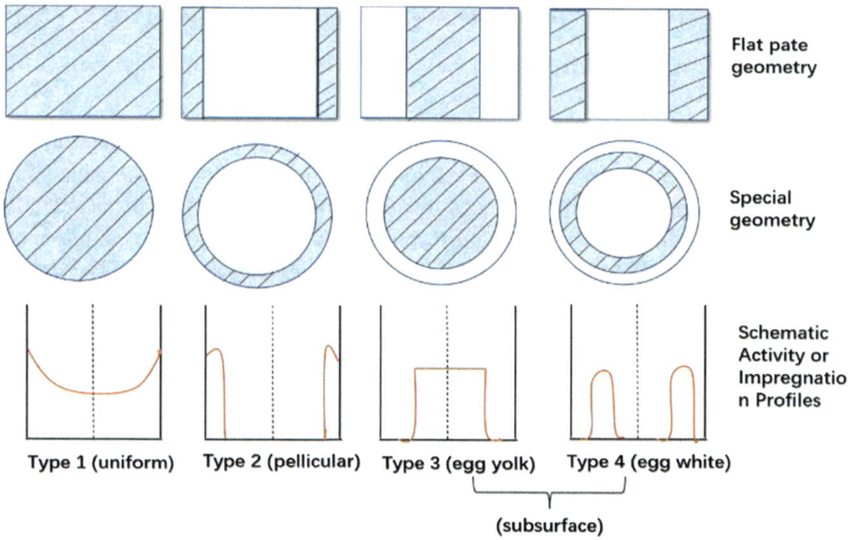

Fig. 3.8 Four limiting types of impregnation or activity profiles. Hatched areas in the cross sections of slabs and pellets (top and middle rows) indicate the regions where active components are deposited. The bottom row shows schematic examples of impregnation or activity profiles

points on the surface of the carrier or deposited, ion-exchanged or even reacted so that the active components are loaded on the carrier, all of which are accompanied by mass transfer processes. After the catalyst is dried, the migration of active components also results as the solvent evaporates. These mass transfer processes do not occur purely and in isolation, but mostly simultaneously and interactively, so the impregnation process must consider the effects of aspiration, deposition, adsorption, and diffusion simultaneously [30].

3.4.2 The Influencing Factors of the Impregnation Method

(1) **The surface properties of the support**

The impregnation process is accompanied by adsorption and the carrier has a certain adsorption capacity for the solutes of the active components. The adsorption capacity of each carrier for the solute of the same active component varies, as does the adsorption capacity of the same carrier for the solute of each active component. However, the solute of an active component can be adsorbed by one carrier under certain conditions (temperature, solvent solids). The curve of adsorption versus solute concentration at a constant temperature is called the adsorption isotherm. It is based on the change in solute concentration in the solution before and after the adsorption has reached equilibrium [31].

(2) **Impregnation time**

The impregnation time is an important parameter and a key factor for the distribution of the active components. In production, the impregnation time is determined based on laboratory work, examining the effect of the impregnation time on the loading site of the active component on the carrier and the concentration distribution of the active component within the catalyst particles, and finally on the catalytic performance of the resulting catalyst. In general, most catalysts require a homogeneous distribution of active components and a high degree of dispersion of the metal particles after reduction to optimize the catalyst activity. The impregnation time is greater than the diffusion impregnation time so that the active components are evenly distributed. At the same time, the production cycle should be taken into account, and for some catalysts, the impregnation time is too long and can be adjusted using competing adsorbents [29].

(3) **Concentration of impregnating solution**

The concentration of the impregnating solution is theoretically estimated according to the requirements of the formulation, the active component content of the catalyst, the pore volume of the carrier, the liquid-to-solid ratio during impregnation, and the adsorption of the active component by the carrier. When the catalyst requires a highly active component content, a highly concentrated impregnating solution is required. However, the concentration of the impregnating solution is too high to penetrate the micro-pores of the granular carrier, resulting in uneven metal loading inside and outside the carrier particles and blocking the micro-pores of the carrier, which is more likely to occur during drying; the high concentration of the impregnating solution is limited by the solubility of the compound, some of which will be close to saturation and need to be heated to dissolve the metal salts of the active component [32]. In the preparation of metal-loaded catalysts, the use of high concentrations of impregnating solution tends to produce coarser metal grains and widen the size distribution of metal grains in the catalyst. To overcome these disadvantages, low-concentration impregnation solutions are often used [33].

3.4.3 *Brief Conclusion*

The sol–gel method has become a common preparation method for NH_3-SCR catalysts due to its simplicity of preparation, high reproducibility, and the possibility of mass production of catalysts. However, a drawback to this method is sintering, particularly for catalysts with high oxide loading. In the wet impregnation method, metals are dispersed only on the surface because the precursor is majorly distributed on support. Consequently, the procedure is determined by the high utilization and low dosage of the active metal, which means that catalysts from this method may not be uniformly spread [34]. Finally, we also need to discuss the relationship between drying and impregnation. The final metal distribution for weak or moderate adsorption is mainly influenced by drying conditions. Strong adsorption, on the other hand, clarifies the effect of impregnation on the final metal distribution. This depends on the ratio of the amount of metal dissolved in the solvent during impregnation to the amount of metal deposited on the support. This is an aspect that must be given full attention during the preparation of NH_3-SCR catalysts by the sol–gel method.

3.5 Hydrothermal Method

The hydrothermal method is a promising liquid-phase preparation technology that has developed rapidly in recent years for the NH_3-SCR preparation, which has resulted in many new methods during the long-term research process, such as adding other force fields to the hydrothermal condition reaction system. These force fields mainly include direct current, electric, magnetic, and microwave fields.

3.5.1 *Basic Concepts of Hydrothermal Method*

The hydrothermal method is a type of soft chemical synthesis method developed by simulating the formation process of some ores in nature. It can be used to grow a

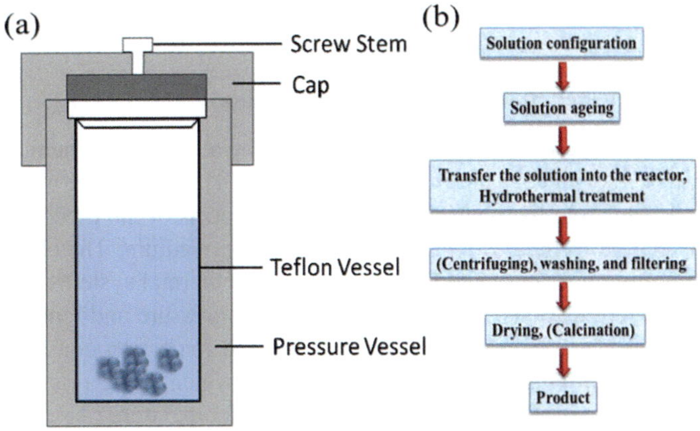

Fig. 3.9 **a** Schematic of typical hydrothermal method equipment. **b** General steps for hydrothermal preparation

variety of single crystals to prepare ultrafine agglomerated or less agglomerated crystallized ceramic powders to complete certain organic reactions. The hydrothermal method refers to the use of an aqueous solution as a reaction system in a special closed reaction vessel (Fig. 3.9a) to create a high-temperature, high-pressure reaction environment by heating the reaction system and pressurizing it (or the vapor pressure generated by itself) [35]. The process dissolves and recrystallizes a substance that is poorly soluble or insoluble under normal conditions [36, 37], and the general preparation steps of the hydrothermal method are shown in Fig. 3.9b.

3.5.2 *Principles of Hydrothermal Synthesis Methods*

(1) **Reaction kinetics and the crystal growth mechanism of the hydrothermal method**

Under hydrothermal circumstances, the primary stages of crystallization are as follows: First, to enter the solution, the reactants should be dissolved in a hydrothermal medium before they may take the form of ions or molecular groups. Second, the difference in temperature between the upper and bottom parts of the kettle separates the ions or molecules. The seed crystal is developed to create a supersaturated solution once the ions or molecular groups are delivered to the low-temperature area. Third, at the growth contact, the ions or molecular groups are adsorbed, broken down, and desorbed. Fourthly, the interface is where the adsorbed material is moving. The dissolved material eventually crystallizes. The growth circumstances have a direct impact on the crystal shape of hydrothermal crystals [38, 39]. The same crystals may show different morphologies under different hydrothermal conditions. It is very important to study the crystal morphology to predict the crystal growth mechanism [40–42].

(2) **The role of water in the Hydrothermal Method**

In a hydrothermal reaction, water can participate in the reaction as a chemical component or it can be a solvent or a puffing accelerator [43, 44]. By speeding up the osmotic reaction and regulating the physical and chemical aspects of the process, inorganic compounds can be formed as a pressure transmission medium. The characteristics of water in high-temperature and high-pressure hydrothermal systems will result in the following modifications: (1) With an increase in pressure and temperature, the ionic product of water rises quickly. As a result, the ionization constant of water rises as the reaction temperature and pressure rise, which is the primary cause of the increase in the hydrothermal reaction. (2) As the temperature rises, the viscosity and surface tension of water decrease. Crystals form more quickly in hydrothermal systems than in other environments because the viscosity of water is reduced and the mobility of molecules and ions in solution significantly increases. (3) The dielectric constant often drops with rising temperature and rises with rising pressure, thus the dielectric constant is frequently low. The reaction is mostly influenced by temperature in hydrothermal settings, and the dielectric constant of water is dramatically decreased. This decline has an impact on water's solvent properties and behaviour. Basic classification of hydrothermal reactions.

There are few hydrothermal studies regarding non-aqueous solvent systems. Most of the materials prepared using the hydrothermal method are oxides and oxygenated salts. Therefore, research regarding the physicochemical properties of hydrothermal devices, solvents, and mineralizers, as well as the study of the mechanisms of chemical reactions during hydrothermal processes, are of great significance in promoting the use of hydrothermal methods.

3.5.3 *Brief Conclusion*

Finally, an introduction to the classification of hydrothermal methods for NH_3-SCR catalysts preparation. According to the research object and purpose, the hydrothermal method can be divided into hydrothermal crystal growth, synthesis, reaction, and treatment. According to the reaction temperature, it can be divided into a low-temperature hydrothermal method and a supercritical hydrothermal method. According to the equipment differences, it can be divided into the ordinary hydrothermal method and a special hydrothermal method. The so-called special hydrothermal method refers to the addition of other action fields to the hydrothermal reaction condition system, such as direct current electric, magnetic, and microwave fields. With the development of modern materials science and engineering research, the application field and basic theory of hydrothermal methods will be further developed. Simultaneously, with the intersection of disciplines, the combination of the hydrothermal method with other methods will be a developmental trend, which will result in the hydrothermal method being more widely used.

References

1. Schwarz JA, Contescu C, Contescu A. Methods for preparation of catalytic materials. Chem Rev. 1995;95:477–510.
2. Ertl G, Knözinger H, Weitkamp J. Preparation of solid catalysts. John Wiley & Sons; 2008.
3. Xu B, Zhang Q, Yuan S, et al. Morphology control and characterization of broom-like porous CeO_2. Chem Eng J. 2015;260:126–32.
4. Kashchiev D. Nucleation. Elsevier; 2000.
5. Kashchiev D, Van Rosmalen G. Nucleation in solutions revisited. Cryst Res Technol: J Exp Indus Crystallogr. 2003;38:555–74.
6. Veesam SK, Ravipati S, Punnathanam SN. Recent advances in thermodynamics and nucleation of gas hydrates using molecular modeling. Curr Opin Chem Eng. 2019;23:14–20.
7. Tromp RM, Hannon JB. Thermodynamics of nucleation and growth. Surf Rev Lett. 2002;9:1565–93.
8. Hoare MR, Pal P. Physical cluster mechanics: statistical thermodynamics and nucleation theory for monatomic systems. Adv Phys. 1975;24:645–78.
9. Wang C, Yang G. Thermodynamics of metastable phase nucleation at the nanoscale. Mater Sci Eng R Rep. 2005;49:157–202.
10. Buell AK. The nucleation of protein aggregates-from crystals to amyloid fibrils. Int Rev Cell Mol Biol. 2017;329:187–226.
11. Lacmann R, Herden A, Mayer C. Kinetics of nucleation and crystal growth. Chem Eng Technol. 1999;22:279–89.
12. Garside J. Nucleation. In: Biological mineralization and demineralization. Springer; 1982. pp. 23–35.
13. Nývlt J. Kinetics of nucleation in solutions. J Cryst Growth. 1968;3:377–83.
14. Garside J, Davey RJ. Invited review secondary contact nucleation: kinetics, growth and scale-up. Chem Eng Commun. 1980;4:393–424.
15. Andrade ÂL, Souza DM, Pereira MC, et al. PH effect on the synthesis of magnetite nanoparticles by the chemical reduction-precipitation method. Quim Nova. 2010;33:524–7.
16. Yang J, Xu L, Wang Z, et al. Preparation and properties of building decoration ceramic materials from spodumene flotation tailings. Chem Ind Eng Progr. 2020;39:3777–85.
17. Sui R, Charpentier P. Synthesis of metal oxide nanostructures by direct sol–gel chemistry in supercritical fluids. Chem Rev. 2012;112:3057–82.
18. Macwan D, Dave PN, Chaturvedi S. A review on nano-TiO_2 sol–gel type syntheses and its applications. J Mater Sci. 2011;46:3669–86.
19. Lannoy A, Bleta R, Machut C, et al. Block copolymer–cyclodextrin supramolecular assemblies as soft templates for the synthesis of titania materials with controlled crystallinity, porosity and photocatalytic activity. RSC Adv. 2014;4:40061–70.
20. Figueira RB. Hybrid sol–gel coatings for corrosion mitigation: a critical review. Polymers. 2020;12:689.
21. Al-Muhtaseb SA, Ritter JA. Preparation and properties of resorcinol–formaldehyde organic and carbon gels. Adv Mater. 2003;15:101–14.
22. Kocklenberg R, Mathieu B, Blacher S, et al. Texture control of freeze-dried resorcinol–formaldehyde gels. J Non-Cryst Solids. 1998;225:8–13.
23. Job N, Théry A, Pirard R, et al. Carbon aerogels, cryogels and xerogels: influence of the drying method on the textural properties of porous carbon materials. Carbon. 2005;43:2481–94.
24. Ertl G, Knözinger H, Weitkamp J. Handbook of heterogeneous catalysis. VCH Weinheim; 1997.

25. Esposito S. "Traditional" sol-gel chemistry as a powerful tool for the preparation of supported metal and metal oxide catalysts. Materials. 2019;12:668.
26. Guo R-T, Zhou Y, Pan W-G, et al. Effect of preparation methods on the performance of CeO_2/Al_2O_3 catalysts for selective catalytic reduction of NO with NH_3. J Ind Eng Chem. 2013;19:2022–5.
27. Guo RT, Zhen WL, Pan WG, et al. Low-temperature selective catalytic reduction of NO on CeO_2-CuO/Al_2O_3 catalysts prepared by different methods. Environ Technol. 2014;35:1766–72.
28. Gronchi P, Kaddouri A, Centola P, et al. Synthesis of nickel supported catalysts for hydrogen production by sol-gel method. J Sol-Gel Sci Technol. 2003;26:843–6.
29. Lee S-Y, Aris R. The distribution of active ingredients in supported catalysts prepared by impregnation. Catal Rev: Sci Eng. 1985;27:207–340.
30. Komiyama M. Design and preparation of impregnated catalysts. Catal Rev: Sci Eng. 1985;27:341–72.
31. Serp P, Kalck P, Feurer R. Chemical vapor deposition methods for the controlled preparation of supported catalytic materials. Chem Rev. 2002;102:3085–128.
32. Perego C, Villa P. Catalyst preparation methods. Catal Today. 1997;34:281–305.
33. Pizzio L, Caceres C, Blanco M. Acid catalysts prepared by impregnation of tungstophosphoric acid solutions on different supports. Appl Catal A. 1998;167:283–94.
34. Jiang SP. A review of wet impregnation—an alternative method for the fabrication of high performance and nano-structured electrodes of solid oxide fuel cells. Mater Sci Eng, A. 2006;418:199–210.
35. Yang G, Park SJ. Conventional and microwave hydrothermal synthesis and application of functional materials: a review. Materials (Basel). 2019;12.
36. Morey GW, Niggli P. The hydrothermal formation of silicates, a review. J Am Chem Soc. 1913;35:1086–130.
37. Yoshimura M, Suda H. Hydrothermal processing of hydroxyapatite: past, present, and future. In: Hydroxyapatite and related materials. CRC Press; 2017. pp. 45–72.
38. Shejale K, Krishnapriya R, Patil H, et al. Recent advances in ultra-low temperature (sub-zero to 100 °C) synthesis, mechanism and applications of titania (TiO_2) nanoparticles. Mater Adv. 2021.
39. Polsongkram D, Chamninok P, Pukird S, et al. Effect of synthesis conditions on the growth of ZnO nanorods via hydrothermal method. Physica B. 2008;403:3713–7.
40. Donnay JDH, Harker D. A new law of crystal morphology extending the law of Bravais. Am Mineral: J Earth Planet Mater. 1937;22:446–67.
41. Hartman P, Perdok W. On the relations between structure and morphology of crystals. I, Acta Crystallogr. 1955;8:49–52.
42. Davey RJ, Milisavljevic B, Bourne JR. Solvent interactions at crystal surfaces: the kinetic story of alpha.-resorcinol. J Phys Chem. 1988;92:2032–36.
43. Yang G, Park S-J. Conventional and microwave hydrothermal synthesis and application of functional materials: a review. Materials. 2019;12:1177.
44. Fenn P. Nucleation and growth of alkali feldspars from hydrous melts. AGU Spring Meet Abst. 2006:V52A-02.

Chapter 4
Catalyst Characterization

Catalyst characterization is the application of modern physical methods and experimental techniques to study the surface and bulk structures of catalyst materials and correlate them with the properties of catalyst materials to explore the relationship between the macroscopic properties, microscopic structure and catalytic properties of catalyst materials and to deepen the understanding of the nature of catalyst materials.

The characterization of catalysts enables a better understanding of the nature of catalysis, allowing the study of catalytic chemistry to evolve from a process to a science, from macroscopic to microscopic, from phenomenal to essential, from empirical to regular, and from specific to universal. In particular, the development of in situ characterization of catalytic systems has enabled a better understanding of the process of catalytic reactions and the interactions between catalysts and reactants; it is very important for theoretically predicting catalyst properties, solving catalyst design problems and speculating on reaction mechanisms. The fundamental purpose of catalyst characterization is to provide a basis for the design and development of catalysts, to improve existing catalysts or to create new catalysts, to propose new concepts, to discover new laws and to promote the development of theory and applied technology. The common characterization methods of SCR catalysts are summarized in Table 4.1.

4.1 Electron Microscopy

4.1.1 Scanning Electron Microscope (SEM)

SEM has emerged as an essential scientific instrument for investigating the morphology of catalysts at high magnifications (1000 to > 500,000 times) and resolutions better than 1 nm [1]. The synthesis of nanomaterials and developments in

W. Pan and R. Guo, *Low-Temperature Selective Catalytic Reduction Catalysts*,
https://doi.org/10.1007/978-981-97-9298-6_4

Table 4.1 Summary of typical characterization approach for SCR-NH_3 catalysts

Characterization techniques	Aspects characterized or investigated
SEM	Characterize the morphologies of the samples
SEM and TEM	Morphologies and textural features of the catalyst
XRD and TEM	The average size of active particles on the surface of the catalyst
BET	Textural properties of the catalyst (specific surface area, pore volume, and pore size of catalysts)
XRD	Crystalline phases of the supports and catalysts
XPS	Surface chemical states, amount of the active component, and peak intensity of active metal
TPD	Effect of different catalysts surface on SO_2 adsorption
NH_3-TPD	Adsorption state of NH_3 on the surface of catalysts and the acidic sites in the catalyst
TPR	Influence of metal content on the oxidation states of the catalyst
H_2-TPR	Reducibility of the catalysts
Raman	Interaction between confined metal species and support material

chemical engineering have both benefited from SEM's superior resolution to visible-light microscopy. The sample size is constrained by the restricted field of view, as is the case with all microscopy, which makes it difficult to determine morphological characteristics such as average particle size, porosity, and distinctive form. The following describes the operational principle and equipment.

The components of an SEM include an electron source, two or more electron lenses, a deflection system, a vacuum system, a control panel, and a number of detectors that examine the signals produced (Fig. 4.1a). It is necessary to demagnify the beam crossing from the source to a focused point or probe with a spot size of less than 10 nm to produce images. Electromagnetic lenses are used to accomplish this (Fig. 4.1b left panel). The objective lens focuses the electron probe onto the sample surface while the condenser lens demagnifies the beam (an operator-controlled function to focus the image). The stigmator, the scanning coils, and the beam-limiting apertures are all contained within the objective lens. A rectangular raster is produced by the scanning coils, a deflection mechanism that advances the probe across the sample's surface point by point at a preset dwell time [2]. The ratio of the sample raster to the image raster determines the magnification while the scan generator simultaneously generates a raster on the viewing screen. For example, a 10-μm-wide raster on the sample projected at 100 mm on a screen gives a magnification of 10,000 times.

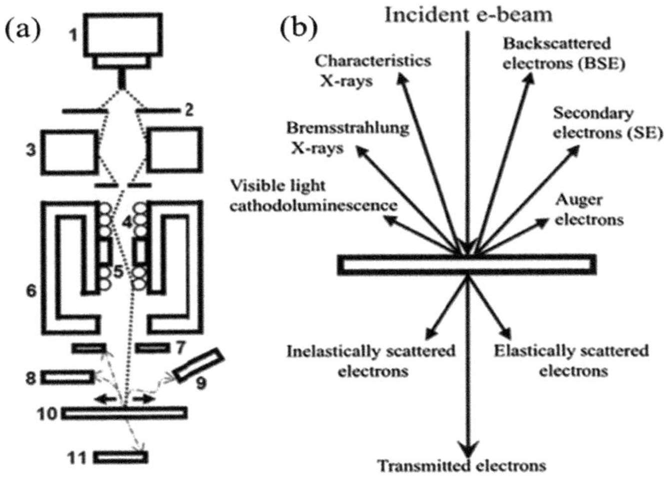

Fig. 4.1 **a** Schematic of scanning electron microscope (SEM) basic components (left panel): (1) electron gun, (2) spray aperture, (3) condenser lens, (4) deflection coils, (5) stigmator, (6) objective lens, (7) backscatter detector, (8) secondary electron detector, (9) X-ray detector, (10) sample stage, and (11) scanning transmission electron detector. **b** Electron beam-specimen interaction (right panel) from Ref. [2]

4.1.2 Transmission Electron Microscope (TEM)

Using a distinct set of magnetic lenses, TEM concentrates the projected electron pattern onto a camera while shaping the incoming electron beam using magnetic lenses onto the object. Apertures placed along the optical path, magnetic lenses, and electron gun parameters are used to regulate the final image's magnification and contrast. With this lens and aperture configuration, TEM can operate in two ways: (1) imaging, which uses high-resolution electron microscopy to reveal the size, shape, crystallinity, and morphology of materials, and (2) diffraction, which uses light rays to analyze the crystal structure of an area of interest on a thin film, particle, or group of particles. For the development of applications relating to energy storage and conversion (catalysis and batteries), many reachers employed TEM as one of their favorite microscopy instruments to research advanced materials such as thin films, nanomaterials, and composites (adsorption, waste-water treatment, and filtration) [3]. The instrumentations and classification are as follows:

(1) **An introduction to the TEM column**

A TEM column contains an electron gun, a series of magnetic lenses and apertures, cameras, and detectors (Fig. 4.2a). Electron beams are generated by tungsten filaments, lanthanum hexaboride crystals, and field emission guns (FEG), however, because FEGs are brighter than other electron beam generators, they are more commonly used in materials examination equipment. Thermionic emission is used by Schottky, or warm FEGs, to produce high current electron sources with 1 eV resolution. Anode plates that drive cold-FEGs have an energy resolution of 0.3 to 0.4 eV and tunnel electrons out of the emitter's tip without raising the emitter's temperature. Since they have a greater energy resolution, they can be used for electron energy-loss spectroscopy (EELS) [4]. The probe's convergence angle and intensity are independently controlled by the condenser aperture and lenses. The probe is larger and parallel in high-resolution electron microscopy (HREM) mode, with a ray path similar to that of traditional optical microscopes. The electron beam condenses to a small probe in scanning transmission electron microscopy (STEM) mode with a significant convergence angle. The objective lens' pole piece generates a strong magnetic field that surrounds the sample. The magnetic field concentrates the electrons on the rear focal plane after the electrons are diffracted and scattered by the material. The objective aperture, also known as the contrast aperture, is positioned on this plane and filters the scattered or diffracted electrons outside of an angle determined by the aperture's size. The image's contrast is improved by removing these electrons. The intermediate picture plane receives a focused image created by the electron beams. This intermediate image is given a selected area aperture to highlight the area that contributes to the final bright field or BF-image.

In diffraction mode, the intermediate and final projector lenses amplify the diffraction pattern generated on the back focal plane, which is similar to increasing the camera length. The intermediate and projector lenses are more intent on bringing the intermediate plane on the camera or viewing screen in BF-TEM, the most popular mode (Fig. 4.2b). For EDX analyses, analytical beryllium or low background holders reduce x-ray absorption Tomography holders tilt up to 70°, compared to the typical 30°. By sandwiching samples between two electron-transparent silicon nitride windows, environmental or in situ holders can work at cold or high temperatures as

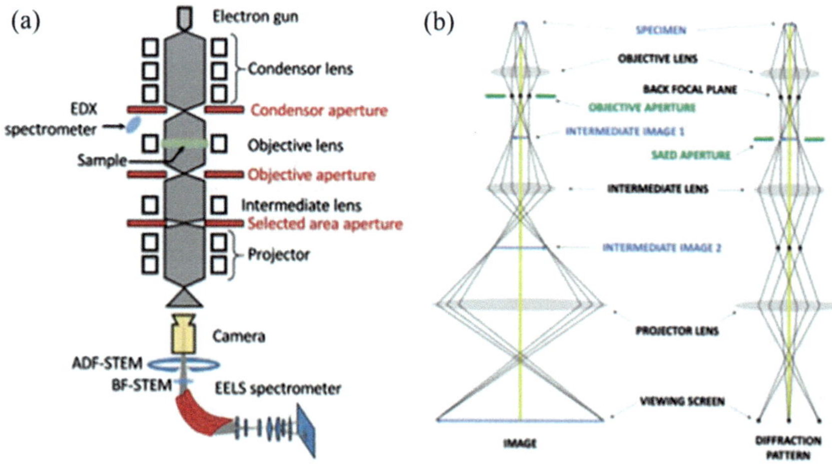

Fig. 4.2 **a** Main components of a TEM column from Ref. [5]. **b** The two modes of a TEM: electron optical ray path when forming an image (left); or a diffraction pattern (right) from Ref. [6]

well as in regulated gaseous environments. This later container is appropriate for operating-microscopy of catalysis [7].

(2) **Scanning transmission electron microscopy (STEM)**

Diffraction-related signals are present in transmitted beam electrons, which lead to artefacts in STEM-BF pictures. With incoherently scattered electrons (scattered above 70 mrad), the diffraction signal can be felted out in STEM mode using an annular detector and create a picture. The contrast becomes a linear function of projected thickness and density in a high-angle annular dark-field (HAADF-STEM), making it simpler to understand. In HAADF mode, the contrast roughly follows a Z^2 relationship (Z is the atomic number) and increases with the scattering power of the atom. For high crystals near the low-index zone axis, Bragg diffracted electrons still exist at these angles, but their number decreases as the angle increases [8].

Spatial resolution in transmission mode is constrained by electron beam scattering over thick materials (> 50 nm for Cu). However, the fundamental factor preventing increased resolution for thinner and more stable materials in the electron beam is the spherical aberration caused by the electron lens [9, 10]. The beam sharpens to below 80 pm using a spherical aberration corrector installed at the condenser level, which is enough to discern the low-index lattices of common metals like Cu111 ~ 208 pm (Fig. 4.3a).

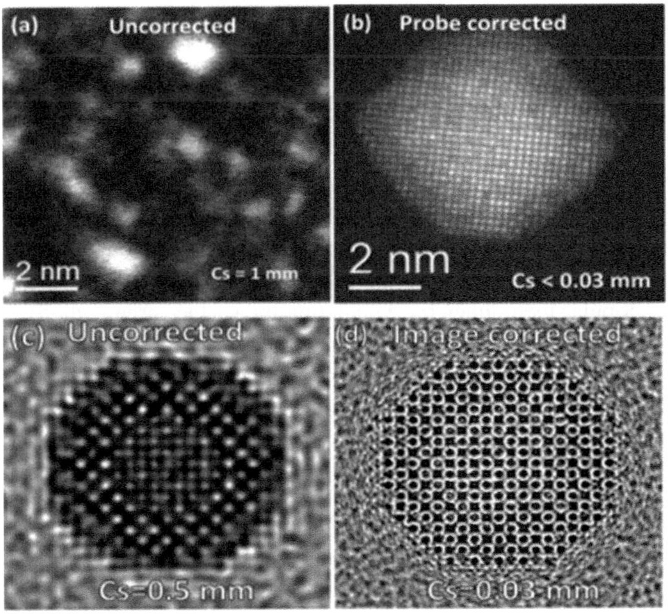

Fig. 4.3 Experimental uncorrected image of metallic nanoparticles using: **a** a conventional STEM-HAADF system; **b** probe-corrected CoPt disordered nanoparticle. Simulated HREM images of a 3.7 nm diameter Co–Pt ordered nanoparticle over amorphous carbon with **c** uncorrected HRTEM uncorrected imaging system (spherical aberration coefficient was set to 0.5 mm); **d** corrected image

(3) **High-resolution electron microscopy (HREM)**

The information limit for HREM mode is the lowest spatial frequency that can be resolved. In this mode, diffracted beams collide to create an image of a crystal's lattice [11]. It is feasible to convert the periodicities of the apparent lattice and the image's symmetry into those of the crystal. Since the information limit in a standard TEM is typically 110 pm, only lattices larger than the Cu planes can be scanned [12].

The darker areas connected to atomic columns on HREM images are not always related to the precise atomic positions. The atomic columns can be seen as dark or brilliant (or completely absent) depending on the local thickness of the crystal and the convergence angle, with the contrast altering with defocus. If the acquisition settings of the HREM are known, the true projected potential of the crystal can be calculated from a through-focus series, or stack of pictures taken at various defocus levels [13]. An example of delocalization where the lattices are projected outside the particle's border is shown in Fig. 4.3a. Due to the disparate projected thicknesses of the center and the edges, the simulation also displays a clear artefact in the particle's center. Additionally, using an image corrector, the ideal Scherzer plane can be brought considerably nearer to the samples. This will decrease the degree of delocalization by bringing a wider range of spatial frequencies that do not experience contrast reversal considerably closer to the sample plane (Fig. 4.3b).

4.2 BET Surface Area

In applications involving porous structures, the specific surface area is the most significant morphological property of solid materials. Micropores are defined as having a pore diameter (d_{pore}) of less than 2 nm, mesopores as having a d_{pore} of between 2 and 50 nm, and macropores as having a dpore greater than 50 nm. Supermicropores ($d_{pore} > 0.9$ nm), medium-sized micropores (0.7 nm d_{pore}), and ultramicropores ($d_{pore} < 0.7$ nm) are three further categories of micropores [14]. One of the most popular techniques for calculating the surface area is the BET method. Zeolites, catalysts, polymers, and photoelectrodes are just a few examples of the numerous kinds of microporous materials that can be characterized using the BET method.

4.2.1 *BET Theory*

The BET equation is: $\frac{p}{V_a(p_0-p)} = \frac{1}{V_m C} + \frac{C-1}{V_m C}\left(\frac{p}{p_0}\right)$. Where P is the pressure, P_0 is the saturation pressure of a substance being adsorbed at the adsorption temperature, C is the BET constant, which is exponentially related to the energy of monolayer adsorption, and V_a is the specific amount of the adsorbed gas at the relative pressure P/P_0. V_m is the monolayer capacity of the adsorbed gas. The form of an isotherm in

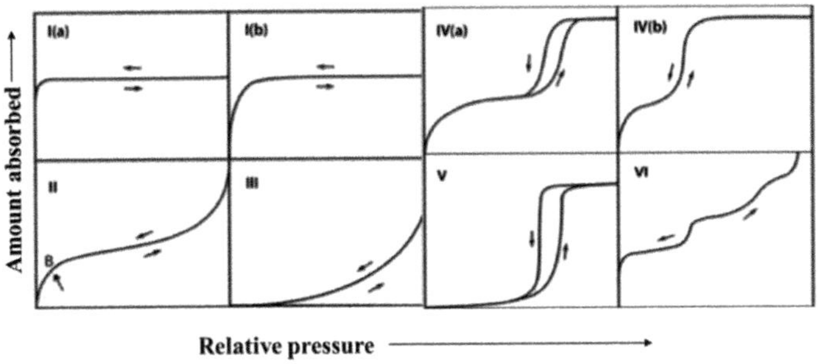

Fig. 4.4 International Union of Pure and Applied Chemistry (IUPAC) classification of BET isotherms from Ref. [17]

the BET range can be determined from the parameter C. The range of applicability of the BET equation explains its limitations. The BET theory's assumption that the adsorbent surface is homogeneous and that there are no interactions between adsorbed molecules is the primary cause of the disparity between theoretical and experimental results. The BET isotherm is still the most common in physisorption research, despite being contentious theoretically.

The fundamental element of BET theory is associated with the adsorption of a gas on the material's surface. Van der Waals forces generated by the adsorbate, which is made up of atoms, ions, or molecules on the surface of a substance that adsorbs these particles, are what lead to this phenomenon. Adsorption can occur through a physical or chemical process. Van der Waals forces are related to physical adsorption, whereas chemical adsorption is the outcome of a chemical reaction between the solid and the adsorbate (gas) [15]. It should be noted that the BET method involves transforming a BET isotherm in order to produce a BET plot. Figure 4.4 shows six different types of BET isotherms [16].

BET isotherms can be characterized as follows: [17]

(a) The reversible type I isotherm, which is typical of microporous solids, has two patterns. Type I(a) is obtained for materials with micropore widths of less than 1 nm. while type I(b) refers to solids that contain both broader micropores and narrow mesopores ($< \approx 2.5$ nm).
(b) Reversible type II isotherm corresponds to materials that are nonporous or microporous. Point B on the isotherm is related to the monolayer coverage. In contrast to a more progressive curvature that denotes the start of multilayer adsorption, an abrupt change in curvature shows that the monolayer coverage has been completed (monolayer coverage overlaps).
(c) Type III isotherm is produced when there are few interactions between the adsorbent and adsorbate. Therefore, information regarding monolayer coverage or creation is unavailable.
(d) Type IV isotherm exists in two types, and they are both correlated with pore width. A type IV(a) is produced if the width is larger than the critical width,

which is determined by the temperature and adsorption properties of the material. Contrarily, a type IV(b) isotherm is frequently seen in mesoporous materials and is reported for materials with mesopores of smaller diameters.

(e) Over low P/P_0 ranges, the shape of a type V isotherm resembles that of a type III isotherm. Weak interactions between the adsorbent and adsorbate are to blame for this phenomenon. At increased relative pressure, hysteresis like in the case of type IV(a) can be observed. Here, pore filling comes after molecular clustering.

(f) Materials with highly uniform nonporous surfaces typically adsorb in multilayers using type VI isotherms. Depending on the substance, the gas, and the temperature, the isotherm takes the form of a stepwise curve.

4.2.2 The Calculations of BET Surface Area

A linear BET range on the plot that has a high regression coefficient must be chosen in order to determine the BET-specific surface area from a BET plot $(P/P_0)/n\ (1 - P/P_0)$ as a function of P/P_0). Typically, pressures between 0.05 and 0.30 are considered typical [17]. However, since the linearity is not always restricted to this range, Rouquerol et al. [18] recommended four consistency criteria as follows:

(a) As a function of P/P_0, the pressure range chosen should monotonically rise with $n\ (1 - P/P_0)$.

(b) The linear regression's parameter C ought to be bigger than zero.

(c) The relative pressure (i.e., P/P_0) for the monolayer loading (i.e., nm) should fall within the chosen linear range.

(d) The relative pressure that corresponds to the calculated value for the creation of the monolayer should be the same as the pressure determined in criteria 3 (although a tolerance of 20% is permitted) [19].

After n_m from the BET equation is obtained, the BET-specific surface area (a_s) can be calculated knowing the average area of value σ_m (molecular cross-sectional area) and implementing the following equation [17].

$$a_s = n_m.L.\sigma_m/m$$

where L is the Avogadro constant and m is the mass of adsorbent.

4.2.3 Drawbacks and Limitations

The BET character is based on gas adsorption, hence monolayers are commonly associated with limits. For illustration, (1) the validity of the BET monolayer capacity is contested, (2) the monolayer structure varies on various surfaces, particularly when N_2 isotherms are used because the molecule is quadrupolar, and (3) strong adsorption at very low-pressure ranges (P/P_0) can entail primary micropore filling in pores with molecular diameters and confined monolayer coverage [20]. When characterizing materials with micropores smaller than 20 nm, micropore filling is often the biggest problem. This could lead to larger or exaggerated surface areas, but it could also reveal significant rates of micropore filling. Studies also show that the BET method may be utilized with microporous materials without encountering problems with micropore filling if the proper adsorption data are used in the right pressure range [21].

4.3 X-Ray Diffraction Techniques

The study of crystal structures and atomic spacing is frequently using X-ray diffraction (XRD), which relies on the constructive interference of monochromatic X-rays with a crystalline sample. The sample is exposed to these X-rays after they have been produced by a cathode ray tube, filtered to produce monochromatic radiation, concentrated by collimation, and directed at it (Fig. 4.5). The distance between these regularly spaced atoms, which is of the same order of magnitude as the wavelength of the incident X-rays, causes the scattered X-rays from the various atoms to interfere with one another and produce strong X-ray diffraction in a few specific directions when a beam of monochromatic X-rays strikes a crystal [22].

4.3.1 Foundations of Crystallography

XRD analysis identifies the long-range order of crystalline materials and the short-range order of non-crystalline materials. We can infer the average grain size, degree of crystallinity, lattice constants, phases, and crystal defects from this data. Advanced XRD provides details on electron density, crystalline symmetry, strain, and roughness. Coherent scattering of the radiation by periodically spaced atoms causes dispersed beams to generate spot patterns from single crystalline samples and ring

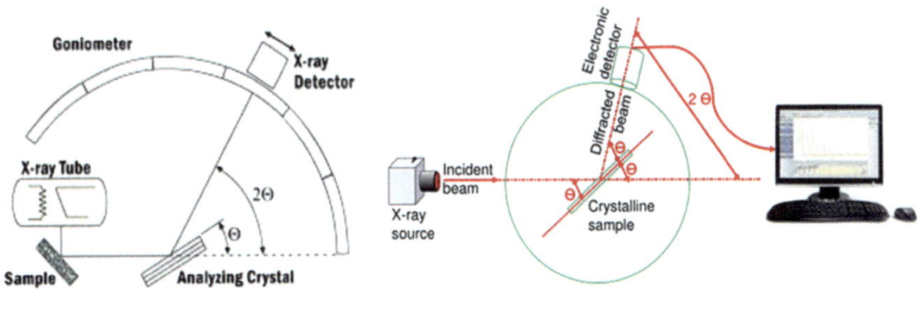

Fig. 4.5 Schematic diagram of a diffractometer system and XRD instrument from Ref. [22]

patterns from polycrystalline materials when radiation impinges upon a solid. A particular crystal structure can be determined by the pattern, diffraction maxima (peaks or lines), and their position (Bragg angle or interplanar spacing d_{hal}) [22]. When a periodic array with long-range order scatters light, it causes diffraction, which creates constructive interference at particular angles.

(1) **X-ray Diffraction for single-crystalline materials**

Physical scattering can be thought of as the incoming wave activating an atom's electrons, which then release radiation with the same wavelength but in the opposite direction of the incoming wave. The wavelength must be of the same order of magnitude as the interatomic distances to produce interference patterns, which is the case for X-rays ($\lambda = 1$ Å). The interaction of the incident rays with the sample produces constructive interference (and a diffracted ray) when conditions satisfy Bragg's Law:

$$n\lambda = 2d \sin \theta$$

n is an integer
λ is the wavelength of the X-rays
d is the interplanar spacing generating the diffraction
θ is the diffraction angle.

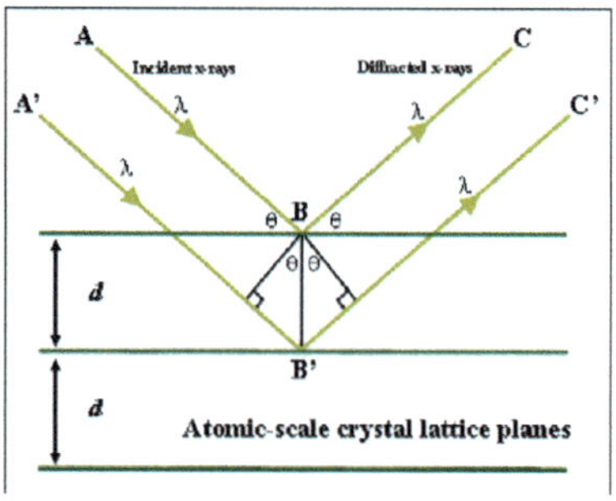

This law establishes a connection between the lattice spacing and diffraction angle in a crystalline sample and the wavelength of electromagnetic radiation. Then, these diffracted X-rays are found, examined, and tallied. Due to the powdered material's random orientation, all potential lattice diffraction directions should be obtained by scanning the sample across a variety of 2 angles. Each compound has a specific set of d-spacings, hence converting the diffraction peaks to d-spacings enables chemical identification. This is often done by comparing the d-spacings to accepted reference patterns [23, 24] (Fig. 4.6).

(2) **X-ray Diffraction for polycrystalline materials**

X-ray polycrystalline diffraction is the definitive method for studying and characterizing the structure of materials. The polycrystalline diffraction spectrum can be used

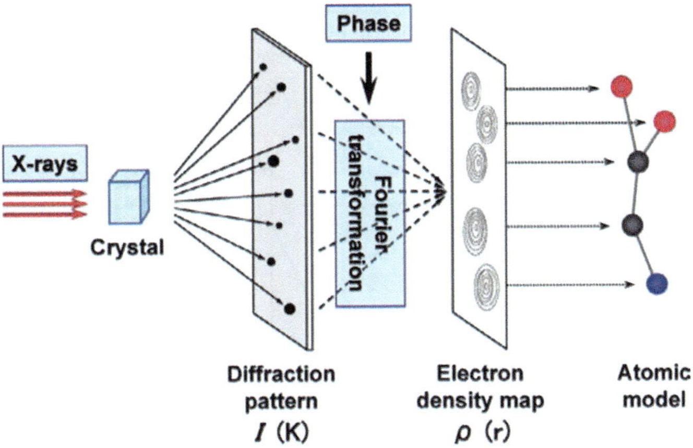

Fig. 4.6 Single crystal XRD reflections from Ref. [25]

to determine a material's distinctive spectrum, which includes elements like peak position, peak intensity ratio, peak form, and half-peak width. The composition of the material phase determines the location and distribution of the diffraction peaks as well as the order of intensities of the peaks. Grain size has a crucial influence on the actual qualities of the material, which also serves as an essential indication for nanomaterials. X-ray polycrystalline diffraction is used to characterize nanomaterials in greater detail. The composition and grain size of nanomaterials in various dispersions, microscopic stresses, preferred orientations, etc., can all be calculated in addition to the grain size [26–28].

4.3.2 Powder XRD Diffraction Analysis

Powder x-ray diffraction has wider applicability compared to SCXRD. The X-ray wavelength is compared to the distance between atoms in PXRD techniques, which show the crystalline structure of materials (Fig. 4.7). The diffraction pattern, which reveals information on the arrangement of the atoms within the crystal, is produced by the scattering of X-rays from atoms. However because they lack a periodic array with long-range order, such as glass and other amorphous materials, their diffraction patterns don't exhibit any prominent peaks [29, 30].

In PXRD, a large number of individual crystallites each intercept a single beam of input x-rays, transforming it into a cone. The detector's spots then produce rings of varying intensities. This method requires that the component has the required absorption coefficient in the absence of a standard. Peak intensities are related to composition, whereas peak positions reveal phase, unit cell features, group spacing, stress–strain analysis, and crystal system. The program, which also identifies crystallographic planes, completes the find-and-match procedure. Peak width and crystal

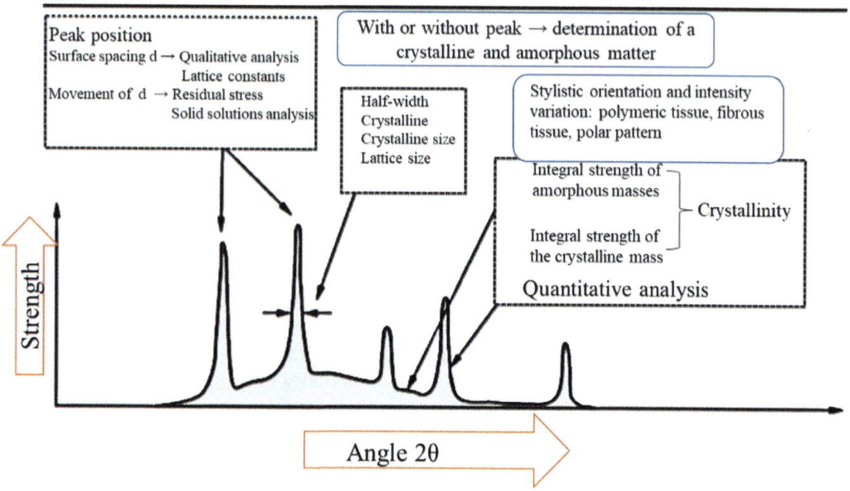

Fig. 4.7 Information obtained from powder XRD profiles

size, lattice distortion, and structural dislocation are all connected (FWHM: full width at half maximum).

According to the Williamson-Hill plot, a straight line can be derived about $\beta.\cos\theta/\lambda$ and $\sin\theta$, with the slope equal to η/λ and the intercept is $1/d_p$. The equation for this line is: $\frac{\beta\cos\theta}{\lambda} = \frac{1}{d_p} + \frac{\eta\sin\theta}{\lambda}$. When $\beta = \text{FWHM}$, d_p is the effective crystallite size, and η is the effective strain. And the crystallite size can be derived from Scherrer's equation: $d_p = 0.9\lambda/\beta\cos\theta$. Additionally, catalysts calcine and crystallize under reaction conditions, increasing the d_p and subsequently the intensity as the peaks shrink. To identify structural changes related to phase transitions and chemical reactions, in situ XRD detects changes in catalyst and other material structures during synthesis and reaction [31].

4.3.3 Application of X-Ray Diffraction in Catalyst Research

Figure 4.8 is a summary of the kinds of information about catalytic materials provided by XRD. And we will present a few of them as follows:

(1) **Application in mesoporous molecular sieves**

Mesoporous materials have pore sizes ranging from 2 to 50 nm, and thus their X-ray diffraction is in the small-angle range (0° to 10°). The pore size of various mesoporous materials and the type of pore channel of mesoporous materials (1D laminar, 2D hexagonal pore, 3D continuous pore) can be determined by small angle X-ray diffraction, as shown in Fig. 4.9 [33].

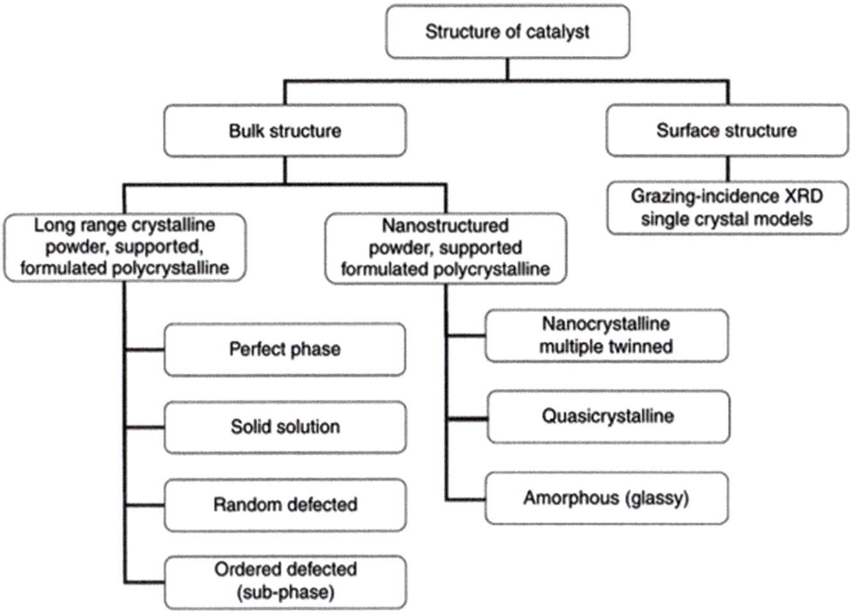

Fig. 4.8 Hierarchy of structural information obtained from XRD characterizing catalytic materials from Ref. [32]

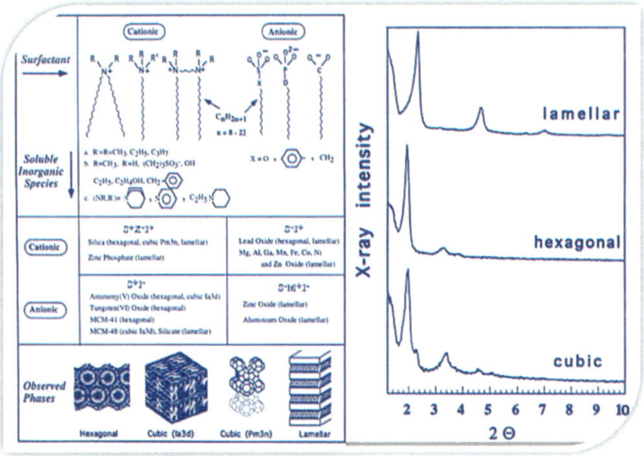

Fig. 4.9 Generalized approach for the synthesis of various soluble inorganic species and surfactant species, powder X-ray diffraction (XRD) patterns of mesostructured antimony oxide materials from Ref. [33]

The discovery of this new class of materials, mesoporous molecular sieves, posed several challenges to our understanding of the formation of porous materials. Numerous observations suggested a close relationship to the surfactant behavior, which was amply exploited in the following exploration and extension of the initial work, but the details had to be worked out. For example, the Si NMR had indicated predominantly amorphous walls and continuous variation of hexagonal lattice constant suggested that XRD patterns could be generated by the order of the pores and not the crystalline structure, which was another key feature of this new class of porous materials [34].

(2) **Applications in supported/composite catalysts**

For some strongly exothermic reactions, the carrier dilutes the active components of the catalyst to meet the thermal equilibrium requirements. Carriers with good thermal conductivity, such as metals and silicon carbide, help to remove the heat of the reaction and avoid local overheating of the catalyst surface. Carriers can also be used to load certain catalysts originally used in homogeneous reactions onto solid carriers to make solid catalysts [35]. XPS can effectively analyses the microstructure, even at low metal contents, and can be used to determine the metal dispersion of metal catalysts. A classic example of XPS is the study of the formation and texture of support materials. Carbon supports were studied in a pioneering set of X-ray experiments both for their similarity with small graphite particles and for their mesoscopic globular structure. This example was chosen not only because it marks the beginning of many studies concerning the secondary structure of graphenes but also because it indicates what kind of information can be obtained with relatively modest experimental effort far below dedicated large-distance detection systems [36].

In all these examples, a common feature is the use of small-angle scattering in combination with various other techniques for obtaining a qualitative or semiquantitative picture of chemical processes relevant to catalysis (Fig. 4.10). The full potential of the method with in situ cells and a deep quantitative analysis as developed in soft condensed matter science and materials science is not frequently used in catalysis. Rapid examination and characterization of composite materials have improved with the availability and precision of XRD apparatus and software. There is still opportunity for technological advancement, nevertheless, to get past constraints on time (analysis takes less than 20 min), limited sensitivity for hybrid materials, hybrid peaks at high angles, and a shared control library of simulated diffraction patterns of different nanomaterials. Software improvements alleviate some of these restrictions so that users can benefit more from the amount of data in XRD spectra.

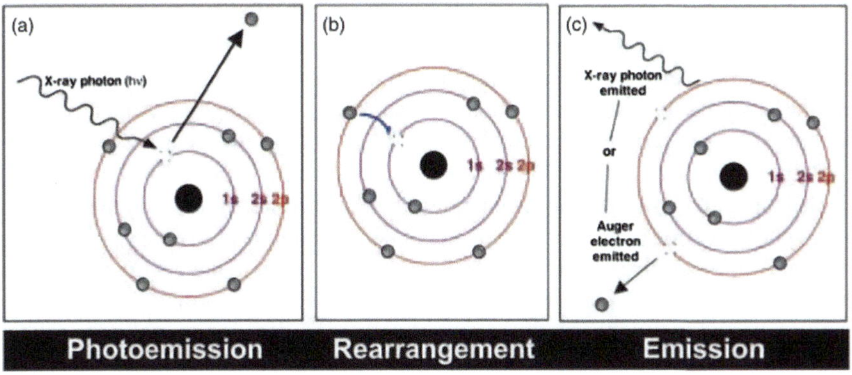

Fig. 4.10 **a** The X-ray photon transfers its energy to a core-level electron, causing photoemission from the starting state of the n-electron. **b** The atom, now in a (n − 1)-electron state, can reorganize by transferring one electron from a higher energy level to the open core hole. **c** Because the electron in (b) dropped to a lower energy level, the atom can expel an electron from a higher energy level to get rid of the extra energy. An Auger electron is the name given to this released electron. The atom can potentially release energy by producing an X-ray photon, a process known as X-ray fluorescence from Ref. [37]

4.4 X-Ray Photoelectron Spectroscopy Techniques

The high information content, adaptability to a wide range of samples, and solid theoretical foundation of XPS as a surface examination method are credited for their popularity. The basis for XPS is the photoelectric effect, which occurs when high-energy photons (often in the keV range) impact material and cause the emission of electrons (photoelectrons). The photoelectron kinetic energy, E_k, which is the measured quantity in the experiment, is given by Einstein's law:

$$E_k = h\nu - E_b \tag{4.1}$$

where E_b is the binding energy of an electron at a specific level and $h\nu$ is the energy of incident radiation. If the incident photon has enough energy, it may ionize a wide range of levels in the sample. This results in the production of a spectrum that shows all accessible energy levels as a distribution of photoelectrons with kinetic energies determined by Eq. (4.1) [38]. The energy of the Auger transition is given by:

$$E_{KL2L3} = E_K - E_{L2} - E_{L3} - F(L_2L_3; X) + R(L_2L_3) \tag{4.2}$$

When E stands for the levels' binding energies, F $(L_2L_3; X)$ for the energy of the Coulombic interaction between the two holes in the final state, X, and $R(L_2L_3)$ for the relaxation term, which refers to the collapse of the passive electron orbitals in the doubly ionized state. The kinetic energy of the Auger electrons is independent of the incident energy and only relates to the binding energies of the three levels participating in the transition, as indicated in Eq. (4.2). By altering the excitation photon energy, which solely influences the position of the photoelectron peaks, Auger peaks can therefore be separated from photoelectron peaks. Binding and Auger energies both provide details on the chemical surroundings of the emitting atom [39, 40].

Both quantitative and qualitative data are produced by the XPS method. The element specificity of the binding energies and the correlation between the strength of the photoelectron peaks and the element concentration lead to quantitative information. A photoelectron peak's energy location in relation to the energy position of the same level in a reference molecule provides qualitative information.

4.4.1 The Features of XPS Spectra

(1) **Peak widths**

The lifespan of the core hole, instrument resolution, and satellite features all affect the observed breadth of a certain photoelectron peak. The peak width due to the core hole lifetime can be calculated from the Heisenberg uncertainty relationship: $\Gamma = h/\tau$. Where Γ is the intrinsic peak width in eV, h is the Planck constant in eV-seconds, and τ is the core whole lifetime in seconds. For the C1S orbital, Γ is ~0.1 eV. The value Γ is typically larger for inner shell orbitals versus outer shell orbitals for a given element [41].

(2) **Peak fitting**

It is necessary to ascertain the area and E_b of each subpeak for a specific orbital to optimize the information gleaned from XPS spectra. Individual subpeaks are rarely entirely separated from one another in an experimental spectrum since the spacing between subpeaks is typically similar to observed peak widths. To resolve the desired peak parameters, a peak-fitting method must be used. The backdrop, peak form (Gaussian, Lorentzian, asymmetric, or mixes), peak position, peak height, and peak width are among the parameters used in such techniques.

Detailed information can be gleaned from XPS spectra with careful peak fitting. When employing high-resolution spectrometers, it is discovered that five peaks may be precisely fitted to this peak envelope, whereas a C1S spectra of poly (2-chloroethyl methacrylate) was previously resolved into three peak components. High-resolution tools might potentially show vibrational elements that widen the XPS peaks [42].

4.4.2 Case Study

In the survey quantification results, Tables 4.2 and 4.3 relate systematic uncertainties to statistical uncertainties. Calculated from $N^{1/2}$, the statistical uncertainty of mass is quite small. The sample standard deviation S_{mass} derived from the five independent measurements is noticeably higher for the majority of peaks. Systematic uncertainty contributions are responsible for this. The choice of experimental parameters, such as step size while scanning the binding energy, and data analysis by the analyst still leave systematic uncertainty even when the same spot of the same sample is evaluated five times consecutively under the same conditions. When five samples of the same sample are tested (Table 4.2), the sample standard deviation is higher because specimen placement and, presumably, variations within the samples themselves—which happen if the catalyst sample is slightly heterogeneous—introduce additional systematic uncertainty.

XPS analysis uses the binding energy at which photoelectron peaks are found to identify elements or chemical groupings. The area of the peaks is used to determine the elemental composition. Uncertainties are reduced by selecting appropriate experimental parameters and acquiring spectra with a good signal-to-noise ratio. By selecting a suitable quantification model, accuracy increases. With novel ionic species for etching and chemical state imaging, recent commercial XPS systems perform depth profiling on a variety of materials, including organics, with a spatial resolution as low as 3 μm. AgLα is one of the higher energy monochromatic x-ray sources that increases information depth and offers non-destructive depth profiling

Table 4.2 XPS survey quantification results from five consecutive measurements of a single specimen with a mass fraction of 10% of Fe and Co supported on alumina (χ = mass fraction; Δ_{mass} = uncertainty; and S_{mass} = sample standard deviation)

Name	Atomic (%)	χ (%)	Δ_{mass} (%)	S_{mass} (%)	S_{mass} (%) (relative)
Al2p	25.8	33.7	0.19	0.3	0.01
C1S	1.6	0.9	0.02	0.3	0.28
O1s	65.7	51.0	0.11	0.4	0.01
F1s	2.4	2.2	0.02	0.5	0.22
Fe2p3/2	2.1	5.7	0.04	0.3	0.05
Co2p1/2	2.5	6.5	0.06	0.6	0.09

Table 4.3 XPS survey quantification results from the measurement of five different specimens of a sample with a mass fraction of 10% of Fe and Co supported on alumina

Name	Atomic (%)	χ (%)	Δ_{mass} (%)	S_{mass} (%)	S_{mass} (%) (relative)
Al2p	25.3	33.3	0.18	0.6	0.02
C1S	2.0	1.1	0.02	0.5	0.46
O1s	66.1	51.7	0.11	1.1	0.02
F1s	2.4	2.2	0.02	0.3	0.11
Fe2p3/2	2.0	5.4	0.04	0.4	0.07
Co2p1/2	20.2	6.2	0.06	0.8	0.13

of surface layers. The ideal x-ray source for XPS is synchrotron radiation because the monochromatic beam creates a narrower peak and lessens peak overlap.

4.5 Temperature-Programmed Analysis Technique

4.5.1 Temperature-Programmed Desorption (TPD)

TPD was initially presented as a quantitative analytical method for surface characterization of low-surface area samples in ultra-high vacuum (UHV) [43]. It was also demonstrated that this technique has the ability to obtain adsorption energy. The study of high-surface-area catalysts under carrier gas and ambient pressure was later proposed using TPD [44]. Today, a vacuum setting is typically employed for surface science research, and catalysis studies frequently employ combined flow and vacuum setups [45]. As a result, it would be possible to focus subsequently on the important subject: kinetic modelling of the rates of intrinsic gas–solid interactions. The available computer resources enable simulations based on nonlinear regression-based model testing and parameter estimation as well as a thorough physicochemical description of TPD systems. However, the kinetic analysis models should explain the underlying processes that result in the reported rates of desorption [46].

Method to Catalyze Studies The equipment utilized in the temperature-programmed desorption investigations is divided into two sections: (1) a static system for adsorption measurements, (2) a system for the temperature-programmed desorption of the gases adsorbed on the catalyst into the stream of a carrier gas [47].

(1) **The introduction of NH_3-TPD**

The NH_3-TPD method is a dynamic in-situ analysis method that offers details on the catalyst's type of acidic active center, its strength, and the quantity of acid present in the corresponding acid strength. An abbreviated schematic of the NH_3-TPD device is shown in Fig. 4.11. The NH_3 adsorption sample is heated at a predetermined rate of temperature rise after being submerged in an inert gas stream. A sufficient temperature will cause the bonds between the adsorbate and the adsorption center to break, allowing those bound to the acid center to be desorbed. When different active center strengths are present, the adsorbent typically desorbs at various temperatures. A desorption spectrum with one or more peaks is produced as a consequence of monitoring the concentration of the desorbed gas in the outflow stream as a function of the sample temperature. The peak temperature is the temperature that corresponds to each peak's highest value on the graph. The peak temperature that results from a sorbent being desorbed on a catalyst reveals the catalytic material's resistance to that sorbent.

The number of acid sites should be determined by counting the number of base molecules that are chemisorbed on the acid sites. Additionally, measurements of equilibrium constant and enthalpy change of desorption (also known as adsorption

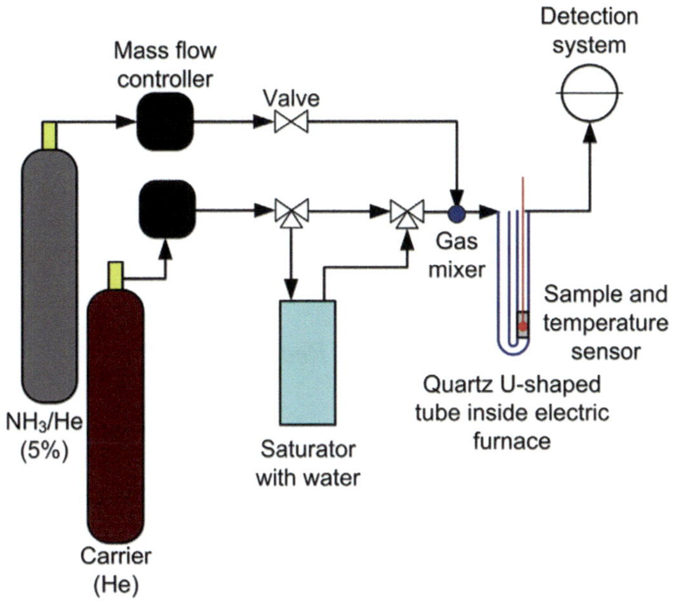

Fig. 4.11 Schematics of an NH_3-TPD experiment at ambient pressure from Ref. [5]

heat) provide a measure of the strength of the intrinsic acid. Methods based on the chemisorption of a gaseous, basic, and tiny molecule are attractive from this standpoint. Microcalorimetry of ammonia adsorption, which is the precise measurement of adsorption heat, is one of the representative techniques. Alkylamines are one way for NH_3-TPD to determine the type of acid site (Brønsted or Lewis) from the perspective of experimental simplicity in achieving repeatable results [48, 49].

(2) **Application of NH_3-TPD in catalysts**

The determination of the peak of desorbed ammonia is necessary for the study of NH_3-TPD. A suitable response is required to the question of the location of ammonia adsorbed. Desorbed ammonia typically shows double or triple peaks [50]. Since the NMR investigation indicated that an excess quantity of NH_3 was hydrogen bonded with NH_4^+ to produce the stabilized $(NH_3)^{n-}$ and NH_4^+, the desorption at low temperatures, such as 400–500 K, is known as the l-peak [51]. The amount of the l-peak is impossible to measure exactly because it is readily desorbed and the amount depends upon the experimental conditions. Ammonia is additionally adsorbed on Na^+ when Na cation is present, and it manifests at the same temperature as when NH_3 on NH_4^+ is desorbed.

Therefore, it is required to enhance the experimental design in order to eliminate the hydrogen-bonded species (l-peak) and collect data regarding the genuine acidic characteristics of solids. Although it is thought that pumping for a long time or at a slightly higher temperature, such as 423 K, is effective, the time/temperature required to completely remove the hydrogen-bonded species cannot be strictly determined for two reasons: (1) the extent of desorption during the pretreatment is influenced by the experimental conditions; (2) the ammonia species hydrogen-bonded, physically adsorbed, adsorbed on a weak acid site may have similar adsorption heats, making it difficult to determine. In addition, Liu et al. [52] successfully predicted the profile of NH_3-TPD using a large ensemble Monte Carlo simulation approach and force field parameters derived from non-initial data from quantum mechanics. This approach provides a means of investigating the relationship between structure and acidity at the molecular level.

4.5.2 *Temperature-Programmed Reduction (TPR)*

TPR is a method for characterizing the chemical composition of solid materials. Numerous articles have been published in the literature demonstrating the use of TPR in numerous scientific disciplines, most notably in the characterization of catalysts [53]. An abbreviated schematic of the H_2-TPR device is shown in Fig. 4.12. The TPR method relies on a species' capacity to be lowered by gas, frequently hydrogen, while the system's temperature is changed in a specific manner (in general, linear temperature increase). The concentration of reducing gas in the effluent gas has reduced relative to its initial concentration, allowing for the monitoring of the reduction process. Information about the species that are present in or on solids can be learned thanks to the outcomes of the research, which is more important than purely analytical information. Most work on the characterization of materials using TPR has been carried out on catalysts and predominantly on supported metal catalysts in the oxidic form [54].

(1) **The introduction of the H_2-TPR**

The temperature-programmed reduction by H_2 (H_2-TPR) has become a ubiquitous technique used to characterize the surface and bulk properties of materials. H_2-TPR is applied to investigate the reduction rates of metal oxide catalysts that reflect the ease of oxygen removal from the catalytic material and are taken to represent the M–O bond strength. In H_2-TPR experiments, a certain amount of metal oxide catalyst is placed in a fixed-bed reactor and a reducing gas stream (usually a mixture of H_2/Ar or H_2/N_2 containing low concentrations of H_2) is passed through the catalyst at a certain flow rate while the catalyst is allowed to warm up linearly at a certain rate. When the temperature reaches a certain value, the oxide on the catalyst starts to be reduced: $MO(s) + H_2(g) \rightarrow M(s) + H_2O(g)$, the change in H_2 concentration after passing through the catalyst bed is proportional to the reduction rate of the catalyst since the reduction gas flow rate is constant [55]. The change in H_2 concentration in the gas stream after passing through the reactor is continuously detected with a gas chromatography thermal conductivity detector and the change in H_2 concentration with temperature is recorded with a recorder, i.e. the TPR spectrum of the catalyst is obtained, which is in the form of a peak-shaped curve. Each TPR peak in the graph generally represents a reducible species in the catalyst and the temperature

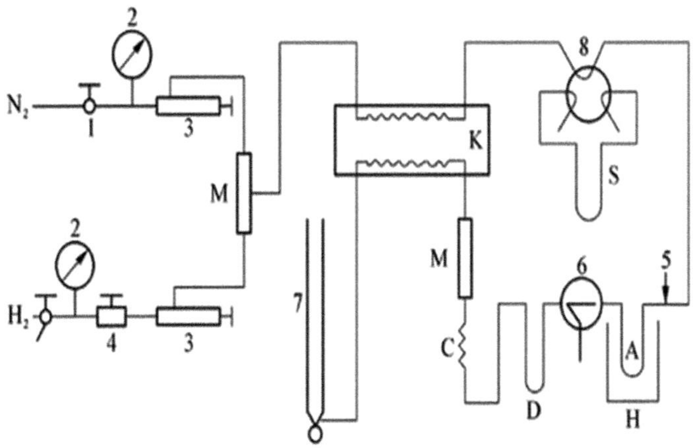

Fig. 4.12 TPR combination units: 1—Pressure stabilisation valves, 2—Pressure gauge, 3—Needle valve, 4—Steady flow valves, 5—Sampler, 6—Three-way valve, 7—Membrane flow meters, 8—Six-way valves; A. Adsorption tubes; C. Coolers; D. Drying tubes; H. Heating furnace; K. Thermal conductivity cell; M. Mixer; S. Quantization tubes

corresponding to its maximum value is called the peak temperature (T_M), the level of T_M reflects the ease of reduction of the oxide on the catalyst and the size of the area under the peak curve is proportional to the amount of the oxide species.

(2) **Application of H_2-TPR in catalysts**

Taking CeO_2 as an example, the characteristic H_2-TPR profile of CeO_2 generally consists of two main H_2 consumption features near 770 and 1100 K. Consumption of H_2 occurring around 770 K has been ascribed to the reduction of surface oxygen species (forming O vacancies), as the total consumption in this region is well-correlated with the surface area of polycrystalline CeO_2 samples [56]. Consumption of H_2 around 1100 K was ascribed to the reduction of bulk CeO_2, resulting from diffusion of O out of the CeO_2 bulk to the CeO_2 surface where it can react with adsorbed or gas phase H species.

4.6 Raman Spectroscopy

Raman spectroscopy has demonstrated its capacity to characterize catalysts and catalytic events at the basic and molecular levels. Raman spectra offer precise details on the structure of the catalysts in the bulk and at the surface, as well as the presence of adsorbates and reaction intermediates, providing significant insight into reaction mechanisms. Modern single-stage spectrometer-based Raman instruments enable great throughput and flexibility in the creation of in situ/operando cells to analyze functioning catalysts.

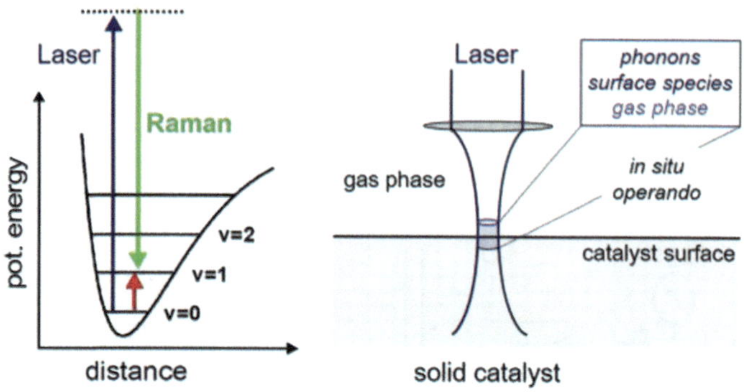

Fig. 4.13 Vibrational Raman spectroscopy of solid catalysts. The Raman signal includes contributions from the surface and the bulk of the solid, whereas gas-phase contributions are usually negligible. Raman spectroscopy may be applied under in situ/operando conditions. The penetration depth of the laser beam depends on the properties of the solid and the excitation wavelength. As shown on the left, the Raman process is based on the inelastic scattering of light. The right figure was modified with permission from Ref. [57]

4.6.1 Basic Principles of Analysis

The studies of Raman spectroscopy mainly focus on gas–solid catalytic reactions, but catalytic reactions in the liquid phase will be touched on if they appear appropriate. Raman spectroscopy is a powerful tool for the in-situ characterization of heterogeneous catalysis as it can provide information about the structure of the catalyst material and surface species (including adsorbates) within one experiment, also under realistic reaction conditions (operando approach), as illustrated in Fig. 4.13. To introduce important features of Raman spectroscopy as applied to catalytic materials, some of the advantages and limitations of the technique will be discussed in the following [57].

Raman spectroscopy's ability to analyze catalyst materials may be constrained by factors such as laser radiation, absorption, fluorescence, and the method's intrinsically low sensitivity. The measurement of spectra is a crucial component of Raman spectroscopy, but it is challenging since the Raman cross sections rely on a variety of factors, including temperature, pressure, the makeup of the gas phase, and excitation wavelength. Relative Raman band intensities are a popular tool for discussing catalytic materials. To date, a wide variety of catalyst materials have been studied using Raman spectroscopy, including bulk oxides, supported metal oxides, supported metal sulphides, molecular sieves, zeolites, bulk metals, supported metals, and heteropoly acids. Here are some examples of applications in catalyst research:

Binary oxides: Solid–solid phase transitions have been characterized (transition order, transition temperature, transition pressure) for many binary oxides, including nanoparticles [58, 59], for example, CeO_2 is a widely used catalyst or support because of its redox properties. Recently, there has been major progress

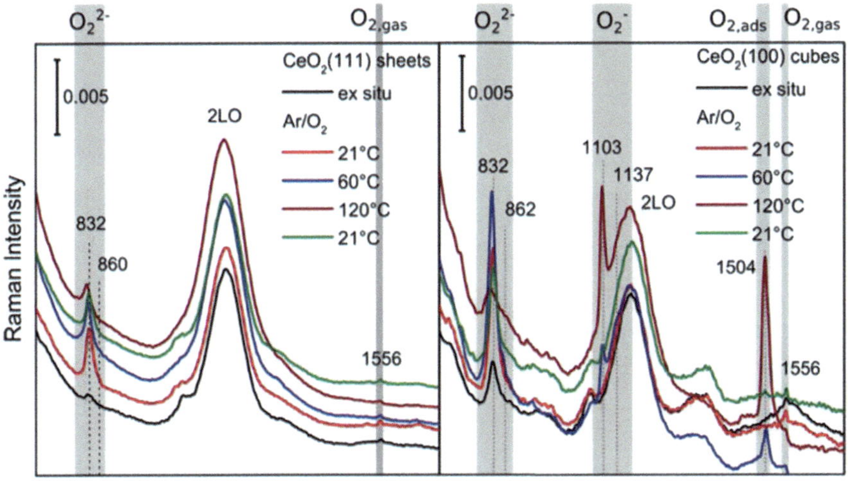

Fig. 4.14 Raman spectra (532 nm) of ceria sheets (exhibiting (111) facets, left) and ceria cubes (exhibiting (100) facets, right), were recorded ex-situ and in situ in 25% O_2/Ar flow at different temperatures. For details see text. Modified from Ref. [60]

in the Raman characterization of the ceria (defect) structure, including the identification of the Ce–O surface modes at around 250 cm^{-1} (longitudinal phonon) and 400 cm^{-1} (transversal phonon), the refined analysis of the profile of the F_{2g} band, and the relation of features in the defect region within 450–600 cm^{-1} to structural motifs (as shown in Fig. 4.14) [60].

Supported metal oxides: Raman spectroscopy has been widely applied to supported metal oxide catalysts, which are relevant for applications and may serve as model catalytic systems [61]. Progress in structural characterization using Raman spectroscopy has been made on supported vanadium oxide systems, using silica, alumina, titania, or ceria as the support, by exploiting UV and visible excitation wavelengths, and by use of visible Raman spectroscopy on titania, bilayered TiO_2–SiO_2, niobium or carbon support materials.

The reaction mechanism of catalysts: the application of Raman MES and PSD has contributed to the mechanistic understanding of the SCR reaction by providing information not accessible by steady-state Raman experiments. To perform MES experiments, the NH_3 reactant was periodically switched on and off, leading to periodic changes in the SCR activity. Under the conditions of the experiment (250 °C, NO steady-state conversion: 26%) the surface of the catalyst was populated mainly by adsorbed NH_3 and water. While the time-resolved spectra do not indicate any significant spectral changes, after PSD, both VO_x and TiO_2 signals responded to the changes in the gas-phase composition (Fig. 4.15a, b), implying the participation of both the VO_x species and the TiO_2 support in the SCR process.

Surface oxygen: Raman spectroscopy allows the detection of different types of surface oxygen features, including lattice oxygen (M–O–M), interface oxygen

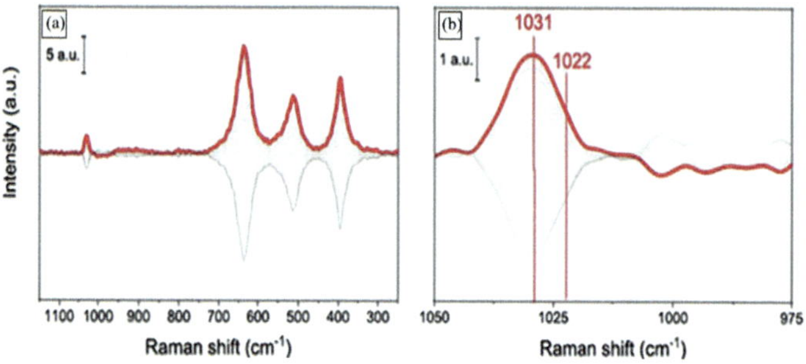

Fig. 4.15 Corresponding phase-resolved spectra of 2 wt% V_2O_5/TiO_2 at 250 °C during 30 s pulses of 500 ppm NH_3 in a gas feed of 500 ppm NO, 2 vol% H_2O, and 5 vol% O_2 balanced in Ar. Modified from Ref. [62].

(M1–O–M2), superoxides (O_2^-), peroxides (O_2^{2-}), and surface hydroxyl groups, as has been demonstrated. For example, ceria surface oxygen (lattice oxygen) features at around 250 and 400 cm^{-1} were identified by combined experimental and theoretical studies [63–65], showing that the 250 cm^{-1} feature originates from O–Ce longitudinal stretching of atoms in the topmost Ce–O layer, and the 400 cm^{-1} feature from the corresponding transversal Ce–O stretching vibration (calculated: 363 cm^{-1}).

References

1. Liu W. Caveats for the use of Web of Science Core Collection in old literature retrieval and historical bibliometric analysis. Technol Forecast Social Change. 2021;172: 121023.
2. Davies TE, Li H, Bessette S et al. Experimental methods in chemical engineering: scanning electron microscopy and X-ray ultra-microscopy—SEM and XuM. Can J Chem Eng. 2022.
3. Braidy N, Béchu A, de Souza TJC, et al. Experimental methods in chemical engineering: transmission electron microscopy—TEM. Can J Chem Eng. 2020;98:628–41.
4. Egerton RF. Electron energy-loss spectroscopy in the electron microscope, Springer Science & Business Media; 2011.
5. Sandoval-Díaz L-E, González-Amaya J-A, Trujillo C-A. General aspects of zeolite acidity characterization. Microporous Mesoporous Mater. 2015;215:229–43.
6. Carter BA, Williams DB, Carter CB et al. Transmission electron microscopy: a textbook for materials science. Diffraction. II, Springer Science & Business Media; 1996.
7. Giorgio S, Sao JS, Nitsche S, et al. Environmental electron microscopy (ETEM) for catalysts with a closed E-cell with carbon windows. Ultramicroscopy. 2006;106:503–7.
8. Pennycook SJ, Nellist PD. Scanning transmission electron microscopy: imaging and analysis. Springer Science & Business Media; 2011.
9. Scherzer O. Über einige fehler von elektronenlinsen. Z Angew Phys. 1936;101:593–603.
10. Scherzer O. Spharische und chromatische korrektur von elektronen-linsen. Optik. 1947;2:114–32.
11. Spence JC. High-resolution electron microscopy. OUP Oxford; 2013.
12. Chen Y, Lu J. Facile fabrication of porous hollow CeO_2 microspheres using polystyrene spheres as templates. J Porous Mater. 2011;19:289–94.

13. Meyer R, Kirkland A, Saxton W. A new method for the determination of the wave aberration function for high resolution TEM: 1 Measurement of the symmetric aberrations. Ultramicroscopy. 2002;92:89–109.
14. Bae Y-S, Yazaydın AO, Snurr RQ. Evaluation of the BET method for determining surface areas of MOFs and zeolites that contain ultra-micropores. Langmuir. 2010;26:5475–483.
15. Garcia-Martinez J, Li K. Mesoporous zeolites: preparation, characterization and applications. Wiley-VCH; 2015.
16. Diwekar U. Batch processing: modeling and design. CRC Press; 2014.
17. Thommes M, Kaneko K, Neimark AV, et al. Physisorption of gases, with special reference to the evaluation of surface area and pore size distribution (IUPAC Technical Report). Pure Appl Chem. 2015;87:1051–69.
18. Catlow CRA, Van Speybroeck V, van Santen R. Modelling and simulation in the science of micro-and meso-porous materials, Elsevier; 2017.
19. Rouquerol J, Rouquerol F, Llewellyn P et al. Adsorption by powders and porous solids: principles, methodology and applications. Academic Press; 2013.
20. Sing K. The use of nitrogen adsorption for the characterisation of porous materials. Colloids Surf A. 2001;187:3–9.
21. Kaneko K, Ishii C. Superhigh surface area determination of microporous solids. Colloids Surf. 1992;67:203–12.
22. Khan H, Yerramilli AS, D'Oliveira A, et al. Experimental methods in chemical engineering: X-ray diffraction spectroscopy—XRD. Can J Chem Eng. 2020;98:1255–66.
23. Causin V, Marega C, Marigo A, et al. Forensic differentiation of paper by X-ray diffraction and infrared spectroscopy. Forensic Sci Int. 2010;197:70–4.
24. Cardell C, Guerra I, Romero-Pastor J, et al. Innovative analytical methodology combining micro-x-ray diffraction, scanning electron microscopy-based mineral maps, and diffuse reflectance infrared Fourier transform spectroscopy to characterize archeological artifacts. Anal Chem. 2009;81:604–11.
25. Hasegawa K. Introduction to single crystal X-ray analysis. Rigaku J. 2012;28:14–8.
26. Welzel U, Ligot J, Lamparter P, et al. Stress analysis of polycrystalline thin films and surface regions by X-ray diffraction. J Appl Crystallogr. 2005;38:1–29.
27. Petkov V. Nanostructure by high-energy X-ray diffraction. Mater Today. 2008;11:28–38.
28. Bernier JV, Suter RM, Rollett AD et al. High-energy x-ray diffraction microscopy in materials science. Annu Rev Mater Res. 2020;50.
29. Dann SE. Reactions and characterization of solids. Royal Society of Chemistry; 2000.
30. Skoog DA, Holler FJ, Crouch SR. Principles of instrumental analysis. Cengage Learning; 2017.
31. Patience GS, Bockrath RE. Butane oxidation process development in a circulating fluidized bed. Appl Catal A. 2010;376:4–12.
32. Schloegl R. X-ray diffraction: a basic tool for characterization of solid catalysts in the working state. Adv Catal. 2009;52:273–338.
33. Huo Q, Margolese DI, Ciesla U, et al. Organization of organic molecules with inorganic molecular species into nanocomposite biphase arrays. Chem Mater. 1994;6:1176–91.
34. Kresge CT, Roth WJ. The discovery of mesoporous molecular sieves from the twenty year perspective. Chem Soc Rev. 2013;42:3663–70.
35. Wang H, Eliaz N, Xiang Z, et al. Early bone apposition in vivo on plasma-sprayed and electrochemically deposited hydroxyapatite coatings on titanium alloy. Biomaterials. 2006;27:4192–203.
36. Brinker CJ, Tallant DR, Roth E et al. Sol-gel transition in simple silicates: III. Structural studies during densification. J Non-Cryst Solids. 1986;82:117–26.
37. Ratner BD, Castner DG. Electron spectroscopy for chemical analysis. Surf Anal Prin Tech. 2009;2:374–81.
38. Carlson T. Photoelectron and Auger spectroscopy. Springer Science & Business Media; 2013.
39. Powell C, Seah MP. Precision, accuracy, and uncertainty in quantitative surface analyses by Auger-electron spectroscopy and x-ray photoelectron spectroscopy. J Vacuum Sci Technol A: Vacuum, Surf Films. 1990;8:735–63.

40. Venezia AM. X-ray photoelectron spectroscopy (XPS) for catalysts characterization. Catal Today. 2003;77:359–70.
41. Bryson CE III. Surface potential control in XPS. Surf Sci. 1987;189:50–8.
42. Jablonski A, Tanuma S, Powell CJ. New universal expression for the electron stopping power for energies between 200 eV and 30 keV. Surf Interface Anal. 2006;38:76–83.
43. Redhead P. Thermal desorption of gases. Vacuum. 1962;12:203–11.
44. Cvetanović R, Amenomiya Y. Application of a temperature-programmed desorption technique to catalyst studies. Elsevier;1967. pp. 103–149.
45. Züchner H, Niemantsverdriet JW. Spectroscopy in catalysis—an introduction. VCH Verlagsgesellschaft, Weinheim, ISBN 3-527-28593-8, 1993, 289 Seiten, Preis DM 148, —, Wiley Online Library, 1994.
46. Cerciello F, Coppola A, Lacovig P, et al. Characterization of surface-oxides on char under periodically changing oxidation/desorption conditions. Renew Sustain Energy Rev. 2021;137: 110453.
47. Cvetanović RJ, Amenomiya Y. Application of a temperature-programmed desorption technique to catalyst studies. 1967. pp. 103–149.
48. Niwa M, Katada N. New method for the temperature-programmed desorption (TPD) of ammonia experiment for characterization of zeolite acidity: a review. Chem Rec. 2013;13:432–55.
49. Čejka J, Wichterlová B. Acid-catalyzed synthesis of mono-and dialkyl benzenes over zeolites: active sites, zeolite topology, and reaction mechanisms. Catal Rev. 2002;44:375–421.
50. Earl WL, Fritz PO, Gibson AA, et al. A solid-state NMR study of acid sites in zeolite Y using ammonia and trimethylamine as probe molecules. J Phys Chem. 1987;91:2091–5.
51. Lónyi F, Valyon J. A TPD and IR study of the surface species formed from ammonia on zeolite H-ZSM-5, H-mordenite and H-beta. Thermochim Acta. 2001;373:53–7.
52. Liu L, Zhao L, Sun H. Simulation of NH_3 temperature-programmed desorption curves using an ab initio force field. J Phys Chem C. 2009;113:16051–7.
53. Hurst N, Gentry SJ, Jones A, McNicol BD. Catal Rev Sci Eng. 1982;24:233.
54. Hurst NW, Gentry SJ, Jones A, et al. Temperature programmed reduction. Catal Rev: Sci Eng. 1982;24:233–309.
55. Kammert J, Moon J, Wu Z. A review of the interactions between ceria and H_2 and the applications to selective hydrogenation of alkynes. Chin J Catal. 2020;41:901–14.
56. Johnson MF, Mooi J. Cerium dioxide crystallite sizes by temperature-programmed reduction. J Catal. 1987;103:502–5.
57. Kim H, Kosuda KM, Van Duyne RP, et al. Resonance Raman and surface-and tip-enhanced Raman spectroscopy methods to study solid catalysts and heterogeneous catalytic reactions. Chem Soc Rev. 2010;39:4820–44.
58. Hakkar S, Achache S, Sanchette F, et al. Characterization by photoluminescence and Raman spectroscopy of the oxide scales grown on the PM2000 at high temperatures. J Mol Eng Mater. 2019;7:1950003.
59. Pagnier T, Boulova M, Sergent N, et al. Nanopowders and nanostructured oxides: phase transitions and surface reactivity. J Raman Spec. 2007;38:756–61.
60. Filtschew A, Hofmann K, Hess C. Ceria and its defect structure: new insights from a combined spectroscopic approach. J Phys Chem C. 2016;120:6694–703.
61. Strunk J, Bañares MA, Wachs IE. Vibrational spectroscopy of oxide overlayers. Top Catal. 2017;60:1577–617.
62. Nuguid RJG, Ferri D, Marberger A, et al. Modulated excitation Raman spectroscopy of V_2O_5/TiO_2: mechanistic insights into the selective catalytic reduction of NO with NH_3. ACS Catal. 2019;9:6814–20.
63. Schilling C, Hofmann A, Hess C, et al. Raman spectra of polycrystalline CeO_2: a density functional theory study. J Phys Chem C. 2017;121:20834–49.
64. Schilling C, Hess C. Real-time observation of the defect dynamics in working Au/CeO_2 catalysts by combined operando Raman/UV–vis spectroscopy. J Phys Chem C. 2018;122:2909–17.

65. Ziemba M, Schilling C, Ganduglia-Pirovano MV, et al. Toward an atomic-level understanding of ceria-based catalysts: when experiment and theory go hand in hand. Acc Chem Res. 2021;54:2884–93.

Chapter 5
MnO_x-Based SCR Catalyst

5.1 Introduction

Currently, W (or Mo)-modified V_2O_5/TiO_2 catalyst has been industrialized for decades [1]. However, vanadium oxide catalysts still have some defects, such as a narrow working temperature range (300–400 °C) and severe biological toxicity [2]. Besides, SO_2 and H_2O could cause the deactivation of V-based catalysts, which also hinder their further application [3]. For solving the mentioned problems, it is crucial to explore a highly-efficient and vanadium-free catalyst to replace V_2O_5-$WO_3(MoO_3)$/TiO_2 for low-temperature NH_3-SCR.

At present, transition metal oxides have received extensive attention as low-temperature SCR catalysts. Among them, manganese oxide-based catalysts exhibit the most excellent low temperature SCR performance due to the following properties: (1) MnO_x-based catalysts have excellent redox properties and strong acidity [4]. Remarkable oxidation–reduction ability is conducive to generating more active NH_2 and NO_2 species and optimizing the reaction pathway of NH_3-SCR [5]. Strong acidity promotes the adsorption and activation of NH_3, which is a prerequisite for the formation of NH_2 species. Furthermore, strong acidity and outstanding oxidation–reduction ability are also two critical factors for the wide temperature window of SCR catalysts. (2) MnO_x-based catalysts with high N_2 selectivity at low temperatures [6], which could avoid secondary pollution caused by the generation of N_2O as much as possible. (3) MnO_x-based catalysts have outstanding SO_2/H_2O resistance at lower temperatures, which could inhibit the formation of sulfate or facilitate the decomposition of sulfate at low temperatures [7]. (4) Compared with noble metal catalysts, MnO_x-based catalysts are competitive in cost and easily available. Consequently, it can be supplied in large quantities.

Typically, MnO_x-based catalysts can be classified into three types: single manganese oxide catalysts, composite manganese oxide-based catalysts, and supported manganese oxide-based catalysts, respectively. Usually, MnO_x-based catalysts are introduced from the following three types: single manganese oxide catalysts,

W. Pan and R. Guo, *Low-Temperature Selective Catalytic Reduction Catalysts*,
https://doi.org/10.1007/978-981-97-9298-6_5

composite manganese oxide-based catalysts, and supported manganese oxide-based catalysts, respectively. In this section, some modification approaches and reaction mechanisms for manganese oxide-based catalysts are also described in detail.

5.2 Single Manganese Oxide Catalysts

Generally, single manganese oxide exhibits promising catalytic performance at low temperatures. However, a series of factors such as manganese valence, crystal structure, specific surface area, surface acidity, morphology, and exposed crystalline surfaces can all affect the activity of single MnO_x low temperature denitration reactions. In this section, the specific effects of these factors on the activity of single MnO_x low temperature SCR are summarized in detail.

5.2.1 Effect of Oxidation State and Crystal Structure on Catalytic Performance

The extranuclear electron arrangement of the Mn atom is [Ar] $4S^2 3d^5$ with five single-electron orbitals. Due to the five single-electron orbitals being in an unsaturated state and the ionization energies at all levels are not very large, it is easier to lose an indefinite number of electrons and have multiple valence states (+ 2, + 3, + 4). The interconversion of Mn^{n+} ions with different valences makes MnO_x possess strong redox performance and surface acidity, which enhances the low-temperature SCR performance of MnO_x-based catalysts. When single-component MnO_x is used as a catalyst in the SCR reaction, the efficiency and N_2 selectivity are closely related to the oxidation state of MnO_x. Compared to Mn_2O_3 and Mn_3O_4, MnO_2 had better denitration performance over a wide temperature range, the NO conversion above 90% at 125–250 °C, which was mainly due to the higher oxidation and better catalytic activity of the high-valence Mn^{4+} ions. Therefore, it is generally believed that MnO_2 has the most exemplary SCR activity in single-component MnO_x. Kapteijn et al. reported the SCR activity of MnO_x with various valence states in detail, and the results proved that the catalytic activity of MnO_x on the unit surface area was in the order: $MnO_2 > Mn_5O_8 > Mn_2O_3 > Mn_3O_4$, further confirms the above point of view.

The crystalline phase of MnO_2 was also one of the essential elements affecting the de-NO_x performance. The [MnO_6] octahedron with the one-dimensional tunnel structure was the basic structural unit of MnO_2 materials. In the [MnO_6] crystal cell, the O atom was located on the corner of the octahedra, and the Mn atom was situated at the center of the octahedra [8]. Multiple units of [MnO_6] octahedrons were connected by co-point, co-edge, or co-planar to form single or double chains. The chains were linked by oxygen atoms at the apex of the octahedron [MnO_6], forming MnO_2 with different tunnels or layered structures. Liu et al. [9] successfully synthesized five

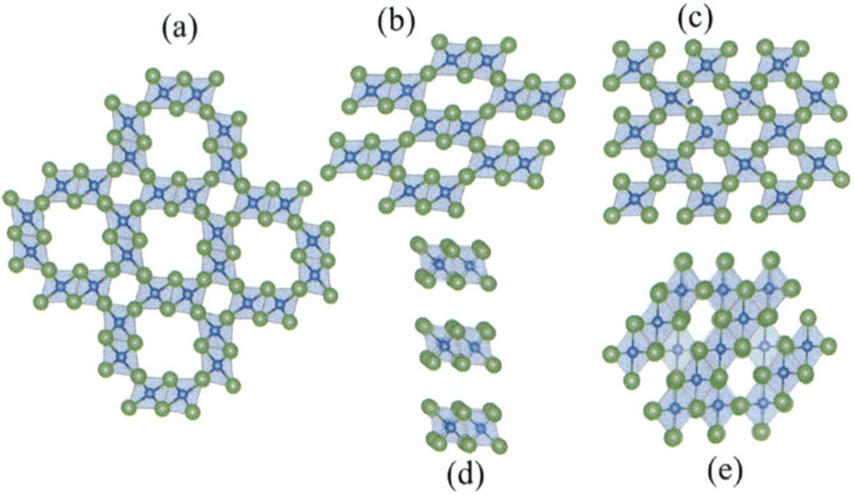

Fig. 5.1 The crystallographic structures of the five kinds of MnO_2: **a** α-MnO_2, **b** γ-MnO_2, **c** β-MnO_2, **d** δ-MnO_2, and **e** λ-MnO_2

kinds of nanocrystalline phase structures of MnO_2, which included α-, γ-, β-, δ- and λ-MnO_2 (Fig. 5.1). α-MnO_2 had a (2 × 2) tunnel structure [10]; γ-MnO_2 had the alternating structure of single (1 × 1) and double (1 × 2) bonds, which was unstable and easy to collapse; β-MnO_2 had a square symmetrical anatase structure, which was relatively stable; δ-MnO_2 was a two-dimensional layered multicrystal structure composed of repeated monolayer MnO_6 lamellae; λ-MnO_2 had a three-dimensional spinel structure with 1 × 1 tunnel [11]. Among the five structures, γ-MnO_2 had the most unstable structure and theoretically had the most excellent low-temperature SCR activity, which was put down to the fact that compared with other crystal structure catalysts, γ-MnO_2 had lower crystallinity, better dispersion, stronger redox performance, more acidic sites, and more abundant chemisorbed oxygen. Gong et al. [12] revealed that the NO_x conversion rates of MnO_2 catalysts with different structures were in the following order: γ-MnO_2 > α-MnO_2 > δ-MnO_2 > β-MnO_2. In the range of 140–200 °C, the NO_x conversion of the γ-MnO_2 and α-MnO_2 catalysts exceeded 90% and the γ-MnO_2 catalyst had the best N_2 selectivity, which further verified the above point of view.

5.2.2 Effect of Specific Surface Area and Surface Acidity on Catalytic Performance

The specific surface area and surface acidity also largely influence the physicochemical characteristics and catalytic activities of single manganese oxide catalysts [13]. The number of active centres in single-phase MnO_x reflected the catalytic activity to a certain extent, and the active sites were evenly distributed on the catalyst surface. Therefore, it was generally agreed that a larger specific surface area could expose more active sites on the surface, increase the contact area between reactants and catalysts, increase the adsorption intensity of NH_3 and NO, and improve the denitration performance of the catalyst [14, 15]. The strong surface acidity is conducive to the adsorption of NH_3, which in turn improves denitration performance. Pretreatment is the most frequently used method to enhance the surface acidity of the single MnO_x catalyst. Jiang et al. [16] treated MnO_2 with acid reflux, the pretreatment increased the content of surface adsorbed oxygen, formed more acid sites, improved the acidic cycle and NH_3 adsorption strength on the MnO_2 catalyst surface, the catalytic activity of MnO_2 was greatly enhanced, and the conversion of NO could exceed 80% in the temperature range of 160–350 °C.

5.2.3 *Effect of Morphology and Exposed Crystalline Surfaces on Catalytic Performance*

The MnO_2 catalyst designed with reasonable morphology could enhance the SCR performance of the catalyst by directly affecting the specific surface area and the content of the active central sites. Compared with bulk particles, manganese oxide materials with rod-like, spherical, nano-flower morphologies displayed better catalytic oxidation performance [17], which was attributed to the fact that the latter had more surface defect sites and lower Mn–O binding strength, increasing the number of active centers, and facilitating the production of more surface adsorbed oxygen (O_α). O_α played an essential role in the SCR reaction, and its migration rate was higher than that of lattice oxygen (O_β), which facilitated the oxidation of NO to NO_2, and the denitration efficiency was improved by the fast-SCR reaction [18, 19]. Adjusting the exposed crystal surface could also improve the denitration efficiency of the catalyst, which increased the content of Mn^{4+} and surface-active oxygen species, providing more active sites [20]. Bai et al. [21] used KIT-6 hard templates to prepare a three-dimensionally ordered mesoporous MnO_2 (3DOM-MnO_2). Compared with the MnO_2 with one-dimensional nanostructure, the three-dimensional structure had a unique mesoporous architecture, bigger specific surface area and more (110) crystalline surfaces exposed on the surface, which facilitated the enrichment of surface Mn^{4+}, thus providing more active sites and improving the catalytic activity. In addition, the exposed crystal planes could also be adjusted by changing the morphologies. When the morphology of α-MnO_2 changed from nanowires to nanoparticles, the primary exposed crystal faces gradually evolved from (200), (310) to (211) [22]. There were more active surface oxygens on the surface of α-MnO_2 (211), resulting in a superior catalytic activity of the nanoparticulate α-MnO_2.

Generally speaking, the activity of MnO_2 catalyst is not determined by a single factor, but the result of the combination of multiple factors. Although it is still uncertain which factor plays the leading role in the catalytic activity of MnO_2, many studies have shown that the changes of acidity, surface active oxygen content and redox property of the catalysts have more influence on the adsorption/activation of NH_3 and NO_x than the crystalline state, specific surface area, and morphology of catalyst. In other words, the chemical properties of MnO_2 can better reflect the SCR performance than the physical properties.

Although the single MnO_x catalyst shows satisfying catalytic performance at low temperatures, there are still many problems in practical application, such as poor thermal stability [23], narrow temperature operation window, poor tolerance to SO_2 and H_2O, low N_2 selectivity and so on [24, 25]. For improving the SCR performance of MnO_x-based catalysts, adding one or several transition metal elements into MnO_x to synthesize the multi-component metal oxide catalyst is the most commonly used method. Research indicated that compared with single MnO_x, the redox properties and surface acidity of composite MnO_x-based catalysts were greatly enhanced [26]. Below we will specifically discuss the measures to improve the de-NO_x performance, N_2 selectivity of pure MnO_x catalyst by doping with one or more metal oxides.

5.3 Multi-metal Manganese Oxide Catalysts

Among the metal elements, titanium, cerium, neodymium, niobium, cobalt, copper, nickel, and several other elements have attracted the most attention as the mixture or dopant to construct binary metal oxide catalysts with MnO_x. Guo et al. [24] pointed out that the Nd-modified $MnNdO_x$-0.1 catalyst exhibited significant catalytic activity and N_2 selectivity in the temperature range of 50–300 °C. In addition, the Nd modification also enhanced the reduction of the catalyst and promoted the adsorption of NH_3 and NO_x along with the presence of more Mn^{4+} and surface adsorbed oxygen species. The results of the in situ DRIFT study indicated that the NH_3-SCR reaction over the $MnNdO_x$-0.1 catalyst was mainly controlled by the Langmuir–Hinshelwood mechanism ($\leq$ 100 °C); the Eley–Rideal mechanism dominates when the reaction temperature exceeds 200 °C (Fig. 5.2). The adsorption of more NO_x and NH_3 species on the $MnNdO_x$-0.1 catalyst, together with the activation effect of Nd on the adsorbed reactants, facilitates the SCR reaction process on the $MnNdO_x$-0.1 catalyst.

In previous research, selectivity and stability of Mn-based catalyst were greatly improved after the modification with Eu [27]. Sun et al. [28] prepared a series of $MnEuO_x$, $MnSmO_x$ and $MnCeO_x$ catalysts by co-precipitation method and used them for the selective catalytic reduction of NO_x by NH_3. It was found that Eu modified MnO_x catalysts with an Eu/Mn molar ratio of 0.1 exhibited the highest NO_x conversion and more excellent N_2 selectivity (Fig. 5.3), which could greatly improve their SCR activity and broaden their operating temperature window, as well as good resistance to SO_2 and H_2O. In addition, the Eu modification resulted in a larger surface area due to the inhibition of crystallization of MnO_x, the enrichment of surface Mn^{4+} and chemisorbed oxygen, and the higher reducibility all contributed to the high SCR activity of $MnEuO_x$-0.1 catalyst. Furthermore, the results of in situ DRIFTS study revealed the NH_3-SCR reactions over MnO_x and $MnEuO_x$-0.1 were

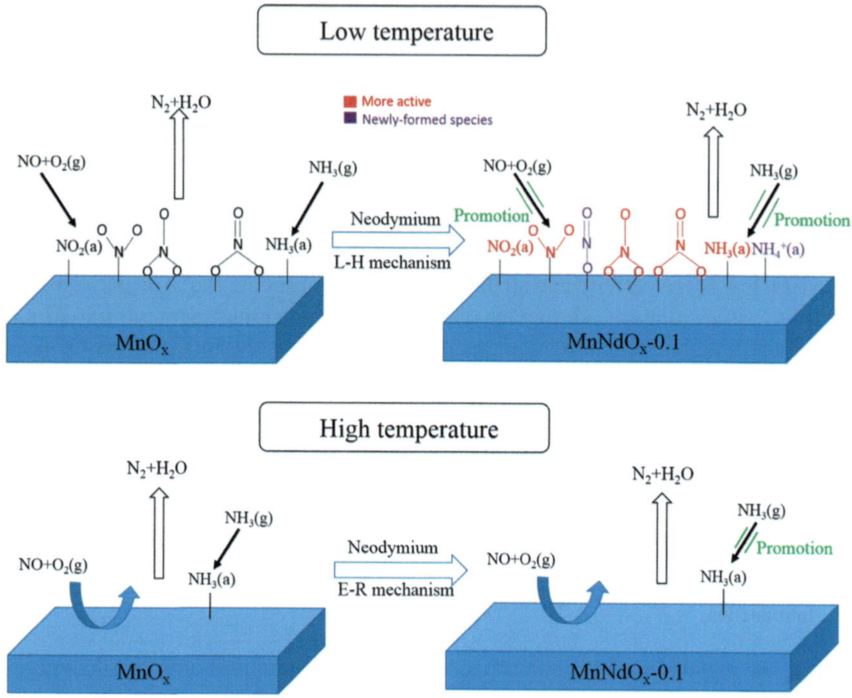

Fig. 5.2 Promotion mechanism of Nd on the SCR reaction over $MnNdO_x$-0.1

mainly controlled by L–H mechanism (< 150 °C) and E–R mechanism (> 150 °C) respectively, and the contribution of E–R mechanism became more important with increasing reaction temperature. The enhanced NO oxidation and NH_3 adsorption should be mainly contribute to the distinguished SCR performance of $MnEuO_x$ catalyst in the lower temperature range (<150 °C) and higher temperature range (> 150 °C) respectively.

Recently, increasing studies showed that the morphology of catalysts played a crucial role in catalytic reactions. Wang et al. [17] synthesized a new spherical catalyst ($CoMnO_x$-BF) for the selective catalytic reduction (SCR) of NO_x by NH_3 by a hydrothermal method. (Fig. 5.4) In addition, a corresponding sample ($CoMnO_x$) was prepared by co-precipitation for comparison. The activity test results show that the $CoMnO_x$-BF catalyst exhibits admirable SCR performance and N_2 selectivity over a wide temperature range of 150–350 °C, together with good SO_2 resistance and high durability. Based on characterization experiments, we can find that the superior surface area resulting from the inhibition of crystallization, the enrichment of surface soluble oxygen and Mn^{4+}, the stronger surface acidity and redox ability all contribute to the NH_3-SCR performance of the $CoMnO_x$-BF catalyst. In addition, enhanced NO oxidation and more active co-reactants on the catalyst surface can also contribute to the excellent SCR activity of $CoMnO_x$-BF at low temperatures.

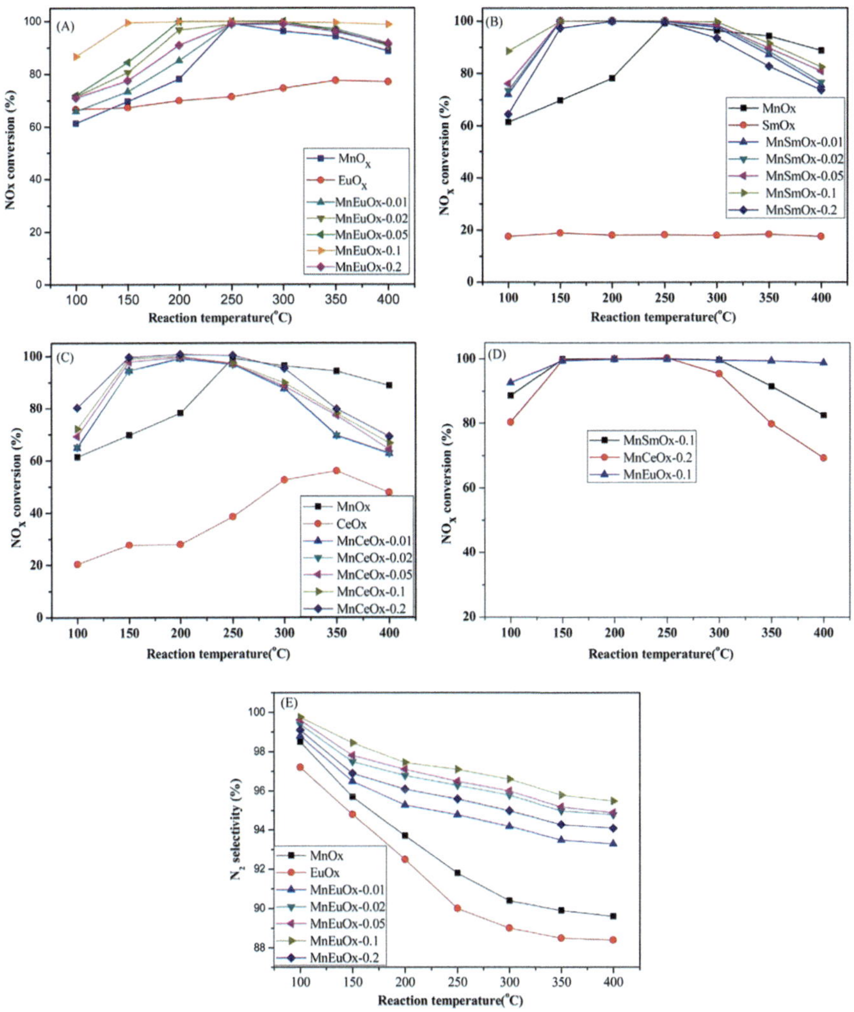

Fig. 5.3 NH_3-SCR performances of MnO_x catalyst modified by different rare earth metals. **a** $MnEuO_x$; **b** $MnSmO_x$; **c** $MnCeO_x$; **d** a comparison among $MnEuO_x$-0.1, $MnSmO_x$-0.1 and $MnCeO_x$-0.2; **e** N_2 selectivity. Reaction conditions: [NO] = [NH_3] = 600 ppm, [O_2] = 5%, balance Ar, GHSV = 108,000 h^{-1}

The morphology of $MnCrO_x$ and $MnCrO_x$-UL catalysts are presented in Fig. 5.5. The $MnCrO_x$ catalyst prepared by hydrothermal reaction exhibits urchin-like morphology, which is built by a spherical core decorated with numerous nanorods, while $MnCrO_x$ contains many agglomerated near-spherical particles. The urchin-like structure of $MnCrO_x$ catalyst could provide more active sites for the contact between catalyst surface and reactants.

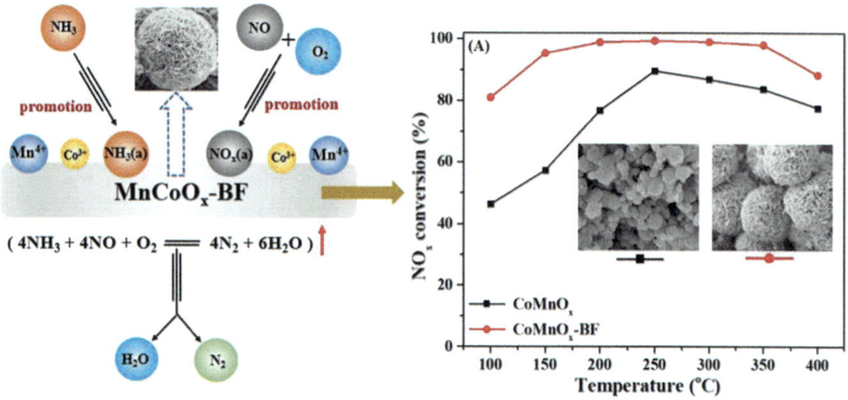

Fig. 5.4 Facilitation of $MnCoO_x$ denitration performance by ball-flowerlike structures

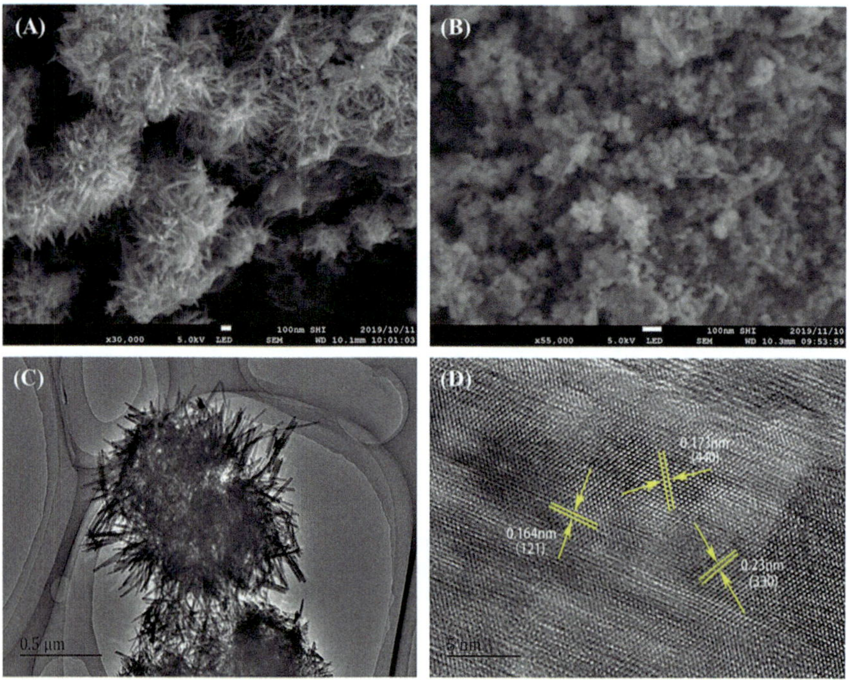

Fig. 5.5 SEM images of **a** $MnCrO_x$-UL, **b** $MnCrO_x$; TEM and HRTEM images of **c**, **d** $MnCrO_x$-UL

TEM micrographs of $MnCrO_x$-UL are showed in Fig. 5.5c, d. From Fig. 5.5c, the detailed urchin-like structure is visible. From the HR-TEM image (Fig. 5.5d), the inter-planar spacing between the adjacent atomic lattice fringes is 0.173 nm and 0.23 nm, which are separately corresponding to the (440) and (330) planes of MnO_2.

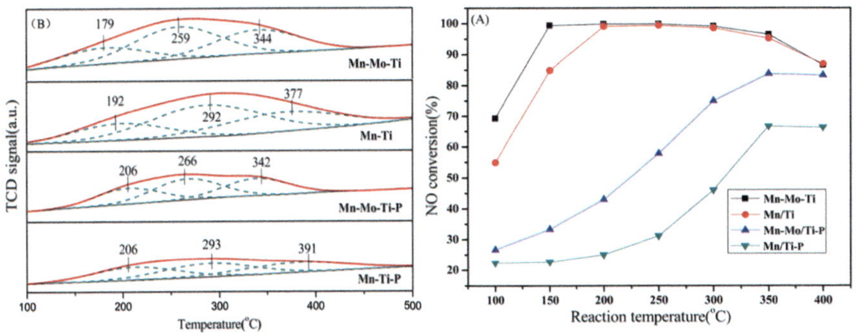

Fig. 5.6 **a** NO_x conversions and **b** NH_3-TPD profiles of the four catalyst samples

Furthermore, the lattice pitch of 0.164 nm should be belonging to the $MnCrO_4$ (121) facet.

Usually, doping the third metal element into the Mn-based composite metal oxide catalyst can further improve SCR performance. Mn/TiO_2 catalyst is a promising candidate for future utilization in low-temperature NH_3-SCR reaction, but it's bad resistance to SO_2 is still a great challenge for practical application. In our previous studies, the main focus was on the modification of Mn/TiO_2 by the addition of a third metal oxide. Guo et al. modified the Mn–Ti catalyst with Mo and tested the catalyst for resistance to P-poisoning. It was found that the addition of Mo greatly improved the NO conversion and tolerance to P species of the Mn–Ti catalyst. NH_3-TPD results revealed that the modification with Mo greatly improved the surface acidity of the fresh catalyst and the catalyst after P poisoning, as reflected in Fig. 5.6. Of these, the first and second peaks can be considered as weak adsorption of NH_3 on the catalyst and the third peak is attributed to strong adsorption of NH_3 on the catalyst. In general, the adsorption of NH_3 species depends largely on the surface acidity of the catalyst, therefore Mo doping greatly increased the amount of NH_3 adsorbed and improved the denitrification performance of the catalyst.

5.4 Supported Manganese Oxide-Based Catalysts

Support is one of the crucial factors affecting the activity of catalysts. Although mixed metal oxides showed more excellent catalytic performance compared to single metal catalysts, their small specific surface area, poor dispersity, and so on all limited their development and application. To further improve the catalytic performance of catalysts, supported metal oxides have been widely studied [29]. The carrier provided active sites for the active component and contributed to the dispersity of the active component, avoiding the degradation of catalytic performance caused by agglomeration. Due to the interaction between the metal components and the carrier, supported metal oxide strengthened the redox performance and the surface acidity of MnO_x-based catalysts, increased the dispersity of MnO_x-based catalysts, and enhanced the electron transfer between the active components. According to the chemical nature of carriers, they could be divided into reducible carriers and irreducible carriers. Reducible supports were usually multivalent metal oxides, such as TiO_2, CeO_2, etc. The reducible carriers themselves provided numerous oxygen vacancies for the catalysts and could sustainably adsorb O_2 in the air to obtain active oxygen species that had a better fluidity and could quickly overflow to the surface of the active components to participate in SCR reactions. Irreducible supports were usually non-variable valent oxides, such as Al_2O_3, SiO_2, etc. The large specific surface area and pore structure of the irreducible carrier contributed to the high dispersal of the active ingredients and greatly improved the activity, stability, and practicality of the catalysts.

5.4.1 *MnO_x-Based Catalysts Supported on TiO_2*

TiO_2 is one of the most extensively applied supports for MnO_x-based catalysts, which can markedly strengthen the denitration performance and anti-SO_2 of the catalysts. The addition of TiO_2 not only promoted the uniform dispersal of the active ingredients, but also provided abundant Lewis acid sites for the adsorption of NH_3.

However, Mn/TiO_2 catalysts also have some drawbacks, such as poor thermal stability and narrow working temperature range. Guo et al. [30] investigated the effect of K and Mg poisoning on the performance of Mn/TiO_2 catalysts. It was found that the addition of K or Mg not only decreased the reducibility of Mn/TiO_2 catalyst but also inhibited the adsorption of reactants. The results of in situ DRIFT study revealed that the introduction of Mg would not change the mechanism of NH_3-SCR reaction over Mn/TiO_2 catalyst, which was a combination of the L–H mechanism (≤ 150 °C) and E–R mechanism (≥ 200 °C), while the NH_3-SCR reaction over Mn/TiO_2–K was mainly controlled by the E−R mechanism in the whole experimental range. For Mn/TiO_2-Mg, the deactivation was mainly due to the inhibited adsorption and oxidation of NO; for Mn/TiO_2–K, the suppressed adsorption and activation of NH_3 species were the primary reasons for its deactivation (Fig. 5.7). Compared to single active component MnO_2/TiO_2 catalysts, multi-active components supported M–Mn/TiO_2 catalysts can largely solve the above problems. Liu et al. [31] prepared Sb-modified MnSb/TiO_2 catalysts by the gel-sol method, and the doping ratio of Sb on the catalysts was recently optimized. Among them, the MnSb/TiO_2-0.25 catalysts showed better SO_2 and H_2O resistance. The XRD patterns of the samples with different Sb doping ratios are shown in Fig. 5.8. The diffraction peaks observed are only rutile and anatase TiO_2, indicating that the Mn and Sb species are in a highly diffuse or amorphous state. The diffraction peaks for the MnSb/TiO_2-0.25 catalyst were the weakest of all the catalyst XRD spectra, which may be a result of the stronger interaction between Mn, Sb and Ti.

The molar ratio of Mn/TiO_2 in multi-active component TiO_2-loaded MnO_x catalysts also largely affects the catalytic performance of MnO_x-based catalysts. Sun

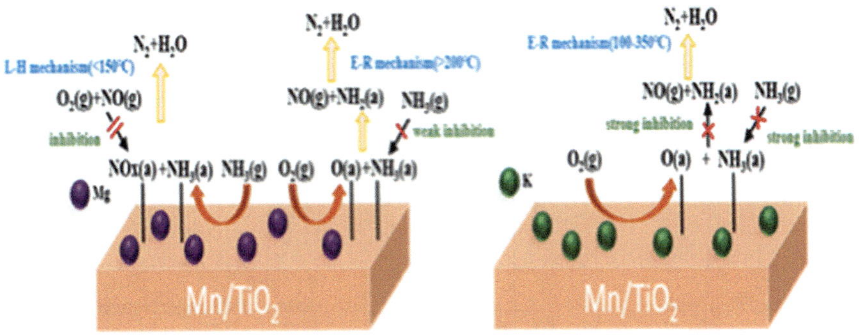

Fig. 5.7 Different poisoning mechanisms of K and Mg over Mn/TiO$_2$ catalysts

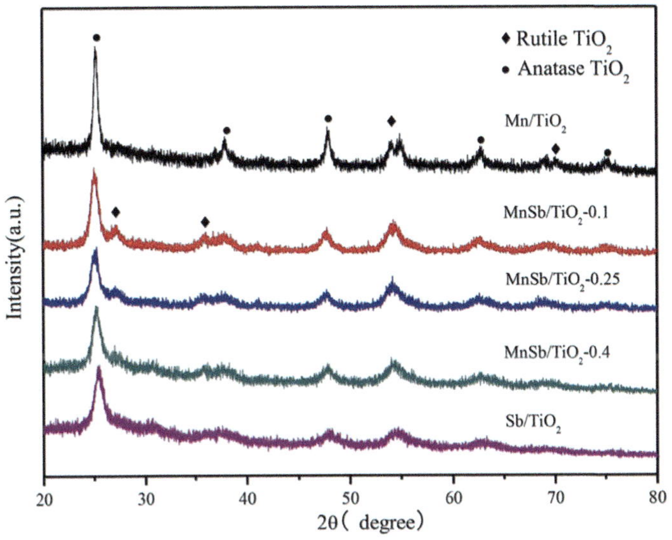

Fig. 5.8 XRD patterns of the catalysts

et al. [5] prepared a series of MnNb/TiO$_2$ catalyst samples with different Nb/Mn molar ratios by sol–gel method. The experimental results showed that the MnNb/TiO$_2$-0.12 catalyst had excellent SCR activity and anti-SO$_2$ performance, and the introduction of Nb into the Mn/TiO$_2$ catalysts could promote the formation of more Mn^{4+} and chemisorbed oxygen on their surfaces, as well as the production of more NO$_2$. Furthermore, the doping of Nb on the Mn/TiO$_2$ catalysts could significantly enhance their surface acidity. The results of the in situ DRIFT study showed that the NH$_3$-SCR reaction over MnNb/TiO$_2$-0.12 catalysts can be carried out via both the Eley–Rideal and Langmuir–Hinshelwood pathways, among which the Eley–Rideal pathway was dominant. In the presence of SO$_2$, the NH$_3$-SCR reaction over Mn/TiO$_2$ and MnNb/TiO$_2$-0.12 catalysts proceed mainly through the reaction between

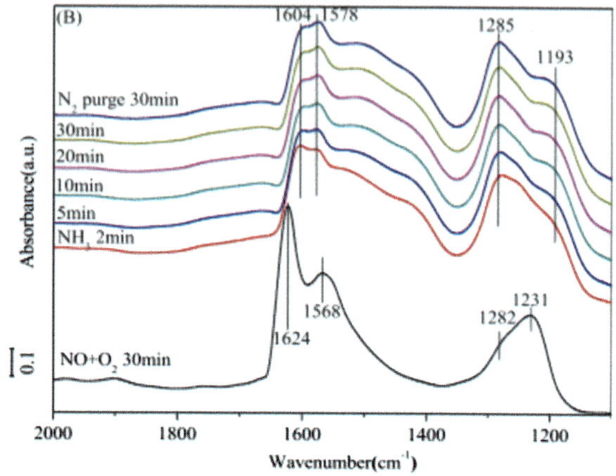

Fig. 5.9 DRIFT spectra of the preadsorbed NO_x species over $MnMo/TiO_2$-0.04 catalysts after the contacting with 600 ppm NH_3 at 50 °C, N_2 balance

ad-NO_x and gaseous NH_3 (Eley–Rideal mechanism), and the generation of more NO_2 over $MnNb/TiO_2$-0.12 catalysts is the major reason for its superior SO_2 tolerance. In addition, Sun et al. [32] synthesized Mn/TiO_2 catalysts with different Mo doping amounts using the co-precipitation method. Among them, the $MnMo/TiO_2$-0.04 catalysts showed the highest SCR activity and the best N_2 selectivity, which were attributed to the decreasing crystallinity, increasing reducibility, increasing surface acid sites, enrichment of surface adsorbed oxygen species and Mn^{4+} species due to the doping of moderate amount of Mo. In situ DRIFT analysis of pre-adsorbed $NO + O_2$ showed that all peaks attributed to ad-NO_x species disappeared and new ad-NH_3 peaks were formed on the catalyst within 2 min after the introduction of NH_3 (Fig. 5.9), indicating that ad-NO_x on the catalyst can react rapidly with ad-NH_3 (NH_3 adsorption is required in the NH_3-SCR reaction). This suggests that the $MnMo/TiO_2$-0.04 catalyst followed mainly the Langmuir–Hinshelwood (L–H).

The poor resistance to alkali metals and heavy metals on Mn/TiO_2 also greatly limits the industrial application of the catalyst. In order to improve the resistance of catalysts to heavy and alkali metals, modification of catalysts with other metals can largely solve these problems. Liu et al. [33] found that the modification of Mn/TiO_2 catalysts with $Al_2(SO_4)_3$ could effectively enhance their Zn resistance. This was attributed to the low crystallinity, high reduction, more acid sites, more Mn^{4+} and chemisorbed oxygen on the $Al_2(SO_4)_3$-modified Mn/TiO_2 catalyst. Furthermore, the results of in situ DRIFTs experiments showed that the addition of $Al_2(SO_4)_3$ did not change the mechanism of the NH_3-SCR reaction on the Mn/TiO_2 catalyst, which followed both Eley–Rideal and Langmuir–Hinshelwood mechanisms. The alkali metals (mainly Na and K) in the fly ash have a serious poisoning effect on SCR catalyst. The loss of reducibility, the decrease of surface acidity and the plug of catalysts channels are usually regarded as the main factors leading to the deactivation of SCR catalyst by alkali metals. It has been reported that Co^{3+} species in cobalt oxides can promote the adsorption of NH_3 and improve SCR activity. Chen et al. [34] prepared Mn/TiO_2 and Mn–Co/TiO_2 catalysts by sol–gel method and the

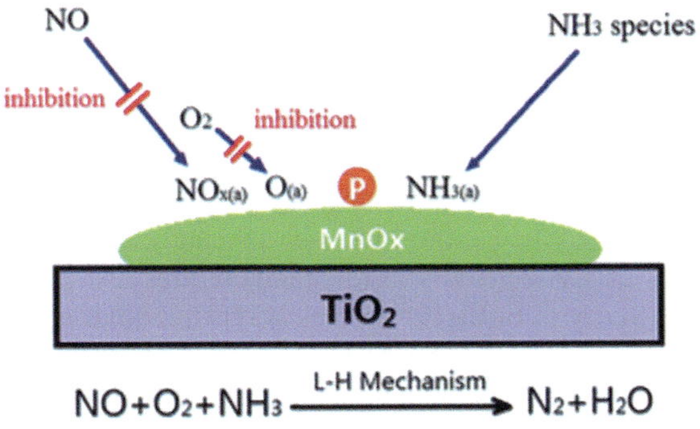

Fig. 5.10 Poisoning mechanism of P addition on Mn/TiO_2 catalysts

results showed that the Co-modified Mn/TiO_2 catalyst can improve the reduction of the catalyst and enhance the adsorption of NH_3 and NO_x over the catalyst. K poisoning had little inhibitory effect on NO_x adsorption over Mn–Co/TiO_2. Therefore, the Mn–Co/TiO_2 catalyst has better K resistance than the Mn/TiO_2 catalyst. Furthermore, Yang et al. [6] investigated the effect of P on the catalytic performance of Mn/TiO_2 catalysts. The results shown that the NH_3-SCR reaction over Mn/TiO_2 catalysts was consistent with the Langmuir–Hinshelwood reaction mechanism. The characterization results revealed that the addition of phosphorus on the Mn/TiO_2 catalyst reduces its specific surface area and reducibility, inhibiting the adsorption of chemisorbed oxygen and NO_x on its surface (Fig. 5.10).

5.4.2 MnO_x-Based Catalysts Supported on Al_2O_3

To widen the temperature range of MnO_x-based catalysts, MnO_x catalysts could be immobilized on the Al_2O_3 support with high porosity and large surface area, which increased the cooperative effect between the active ingredients and the support, facilitated the electron migration. Compared with pure MnO_x, easily reducible MnO_x species could be produced over the MnO_x/γ-Al_2O_3 catalyst, which greatly enhanced the NO conversion in the SCR reaction. Although the single active component MnO_2/Al_2O_3 catalyst had a certain application prospect in low-temperature NH_3-SCR reaction, its weak surface acidity and poor redox performance were still not negligible drawbacks. To upgrade the property of MnO_x/Al_2O_3 catalysts, the modification of Mn/Al_2O_3 catalysts with other metals to construct multi-active component catalysts have been widely studied. Yang et al. [35] introduced Mo into Mn/γ-Al_2O_3 further improved the dispersity of the active components on Mn/γ-Al_2O_3. Secondly, Mo modification could also form more Mn^{3+} on the carrier Al_2O_3 surface, and more NH_3 could be adsorbed on Mn^{n+} to oxidatively dehydrogenate to form $-NH_2$ species, facilitating the progression of reaction. Furthermore, the introduction of Mo also reduced the deactivation of catalyst and enlarged the effective temperature range of SCR reaction. Yang et al. [36] indicated the conversion of NO on Fe–Mn/Al_2O_3

catalyst exceeded 95% at 150–200 °C. This was attributed to the doping of Fe accelerated the electron transfer between Mn^{n+} and Fe^{n+} at different valence states and enhanced the synergistic effect between the active component MnO_x and the carrier Al_2O_3, which improved the redox performance and surface acidity of the catalyst. Although the large pore structure of Al_2O_3 contributed to the high dispersal of the active ingredients, the MnO_x-based catalysts supported by single Al_2O_3 exhibited poor catalytic activity at low temperatures and easily produced $NH_4Al(SO_4)_2$ and $Al_2(SO_4)_3$ to block the stomata [37]. In addition, Huang et al. [38] calculated the electronic structures of Al_2O_3 and TiO_2 models through the density of states (DOS) and found that the band gap width of Al_2O_3 (2.78) was approximately 2.7 times that of TiO_2 (1.03). It was generally accepted that the structures with a small band gap or no band gap could transfer electrons more easily than the structures with large band gap, so the electron transfer capacity of Al_2O_3 was weaker than that of TiO_2, which was not conducive to the catalytic activity of supported catalyst. For improving these shortcomings of Al_2O_3, the Al_2O_3–TiO_2 composite catalyst emerged. Li et al. [39] prepared TiO_2–Al_2O_3 composite support by coprecipitation method, which not only solved the drawback of the easy agglomeration of active components of single TiO_2 supported catalyst, but also greatly improved the surface acidity of single Al_2O_3 supported MnO_x-based catalyst, The obtained $MnCeO_x$/TiO_2–Al_2O_3 catalyst performed more excellent activity and stability than the single Al_2O_3 supported $MnCeO_x/Al_2O_3$ catalyst.

5.4.3 MnO_x-Based Catalysts Supported on Carbon Materials

Carbon materials such as activated carbon (AC) and carbon nanotubes (CNTs) had large surface areas, excellent porosities, good chemical stability, and could provide numerous acid sites in the catalytic process, which were often used as catalyst supports [40]. Carbon materials as supports promoted the uniform dispersal of the active metal components on the catalysts, largely improving the denitration efficiency and N_2 selectivity of the catalysts. Using CNTs materials as supports provided an excellent reaction site for catalytic reaction, which not only enhanced the local concentration of metal particles and internal reactants inside CNTs through a narrow space, but also contributed to enhancing the interplay between the active metal phase and the internal surface of CNTs [41, 42]. The catalyst, obtained by loading MnO_x onto biomass carbon (BC) after nitric acid treatment, exhibited excellent denitrification performance in the temperature range from 125 to 225 °C. This was attributed to the abundant oxygen-containing groups of BC support. Moreover, the graphitic crystallite structure on the surface of BC provided numerous adsorption sites for NH_3 adsorption, and BC could act as the oxidant of NO, which could oxidize NO to NO_2 even at low temperatures [43]. Moreover, most researchers believed that the loading of active metal components on CNTs was closely related to the content of oxygen functional groups on CNTs, and the increase of oxygen functional groups could improve the loading of active metal components on CNTs. Wang et al. [43] treated carbon nanotubes with nitric acid for an extended period of time to increase the number of oxygen functional groups, which greatly improved the loading capacity and dispersion of metal oxides on CNTs carrier, further confirming the above views.

Compared with pure MnO_x catalysts supported on carbon materials, the supported mixed metal oxides exhibited better catalytic activity, which might be due to the synergistic interaction between multi-metals ions and the carrier. Yang et al. [44] prepared Mn–Ce–Fe/AC catalyst by impregnation method. Fe doping not only increased the mobility of oxygen during SCR reaction and increased the adsorption amount of NH_3 on the activated carbon surface by exposing more acid active sites [45], but also increased the average valence state of manganese ions [46]. The redox cycling between the higher valence state of Mn^{n+} and the different valence state of M^{n+} might facilitate the oxidation of NO to NO_2, leading to higher activity at low temperature [47].

References

1. Duan C, Guo R, Wu G, et al. Selective catalytic reduction of NO_x by NH_3 over $CeVO_4$-CeO_2 nanocomposite. Environ Sci Pollut Res Int. 2020;27:22818–28.
2. Zhou J, Guo R, Zhang X, et al. Cerium oxide-based catalysts for low-temperature selective catalytic reduction of NO_x with NH_3: a review. Energy Fuels. 2021;35:2981–98.
3. Duan C, Guo R, Liu Y, et al. Enhancement of potassium resistance of Ce–Ti oxide catalyst for NH_3-SCR reaction by modification with holmium. J Rare Earths. 2022;40:49–56.
4. Guo R, Qin B, Wei L, et al. Recent progress of low-temperature selective catalytic reduction of NO_x with NH_3 over manganese oxide-based catalysts. Phys Chem Chem Phys. 2022;24:6363–82.
5. Sun P, Huang S, Guo R, et al. The enhanced SCR performance and SO_2 resistance of Mn/TiO_2 catalyst by the modification with Nb: a mechanistic study. Appl Surf Sci. 2018;447:479–88.
6. Yang N, Guo R, Wang Q, et al. Deactivation of Mn/TiO_2 catalyst for NH_3-SCR reaction: effect of phosphorous. RSC Adv. 2016;6:11226–32.
7. Li W, Guo R, Wang S, et al. The enhanced Zn resistance of Mn/TiO_2 catalyst for NH_3-SCR reaction by the modification with Nb. Fuel Process Technol. 2016;154:235–42.
8. Yuan Y, Liu C, Byles BW, et al. Ordering heterogeneity of [MnO_6] octahedra in tunnel-structured MnO_2 and its influence on ion storage. Joule. 2019;3:471–84.
9. Liu J, Chen W, Hu X, et al. Effects of MnO_2 crystal structure on the sorption and oxidative reactivity toward thallium(I). Chem Eng J. 2021;416: 127919.
10. Peng Y, Chang H, Dai Y, et al. Structural and surface effect of MnO_2 for low temperature selective catalytic reduction of NO with NH_3. Procedia Environ Sci. 2013;18:384–90.
11. Liu Y, Deng H, Lu Z, et al. The study of MnO_2 with different crystalline structures for U(VI) elimination from aqueous solution. J Mol Liq. 2021;335.
12. Gong P, Xie J, Fang D, et al. Effects of surface physicochemical properties on NH_3-SCR activity of MnO_2 catalysts with different crystal structures. Chin J Catal. 2017;38:1925–34.
13. Liu C, Wang H, Zhang Z, et al. The latest research progress of NH_3-SCR in the SO_2 resistance of the catalyst in low temperatures for selective catalytic reduction of NO_x. Catalysts. 2020;10:1034.
14. Shi J, Gao C, Liu C, et al. Porous MnO_x for low-temperature NH_3-SCR of NO_x: the intrinsic relationship between surface physicochemical property and catalytic activity. J Nanopart Res. 2017;19:194.

15. Feng X, Zhou T, Liu Z, et al. Preparation and low temperature denitrification properties of ferric oxide red supported manganese oxide and manganese cerium co-oxide catalyst. Ceram Int. 2021;47:14988–96.
16. Jiang Y, Cheng G, Yang R, et al. Influence of preparation temperature and acid treatment on the catalytic activity of MnO_2. J Solid State Chem. 2019;272:173–81.
17. Wang Z, Guo R, Shi X, et al. The superior performance of $CoMnO_x$ catalyst with ball-flowerlike structure for low-temperature selective catalytic reduction of NO_x by NH_3. Chem Eng J. 2020;381: 122753.
18. Zhang Z, Li Y, Yang P, et al. Improved NH_3-SCR deNO$_x$ activity and tolerance to H_2O & SO_2 at low temperature over the $Nb_mCu_{0.1-m}Ce_{0.9}O_x$ catalysts: role of acidity by niobium doping. Fuel. 2021;303:121239.
19. Liu S, Guo R, Sun P, et al. Mechanistic investigation of the different poisoning mechanisms of Cl and P on Mn/TiO_2 catalyst for selective catalytic reduction of NO_x with NH_3. J Taiwan Inst Chem Eng. 2017;80:314–25.
20. Zheng X, Cai J, Cao Y, et al. Construction of cross-linked δ-MnO_2 with ultrathin structure for the oxidation of H_2S: structure-activity relationship and kinetics study. Appl Catal B. 2021;297: 120402.
21. Bai B, Qiao Q, Li J, et al. Synthesis of three-dimensional ordered mesoporous MnO_2 and its catalytic performance in formaldehyde oxidation. Chin J Catal. 2016;37:27–31.
22. Pan T, Deng H, Kang S, et al. Facile homogeneous precipitation method to prepare MnO_2 with high performance in catalytic oxidation of ethyl acetate. Chem Eng J. 2021;417: 129246.
23. Meng D, Zhan W, Guo Y, et al. A highly effective catalyst of Sm-Mn mixed oxide for the selective catalytic reduction of NO_x with ammonia: effect of the calcination temperature. J Mol Catal A: Chem. 2016;420:272–81.
24. Guo R, Sun P, Pan W, et al. A highly effective $MnNdO_x$ catalyst for the selective catalytic reduction of NO_x with NH_3. Ind Eng Chem Res. 2017;56:12566–77.
25. Liu J, Guo R, Li M, et al. Enhancement of the SO_2 resistance of Mn/TiO_2 SCR catalyst by Eu modification: a mechanism study. Fuel. 2018;223:385–93.
26. Wei L, Guo R, Zhou J, et al. Chemical deactivation and resistance of Mn-based SCR catalysts for NO_x removal from stationary sources. Fuel. 2022;316.
27. Guo D, Guo R, Duan C, et al. A study on the offset deactivation effect of potassium and phosphorus on $MnEuO_x$ SCR catalyst. ChemCatChem. 2021;13:1743–52.
28. Sun P, Guo R, Liu S, et al. The enhanced performance of MnO_x catalyst for NH_3-SCR reaction by the modification with Eu. Appl Catal A. 2017;531:129–38.
29. Pan W, Hong J, Guo R, et al. Effect of support on the performance of Mn–Cu oxides for low temperature selective catalytic reduction of NO with NH_3. J Ind Eng Chem. 2014;20:2224–7.
30. Guo R, Wang S, Pan W, et al. Different poisoning effects of K and Mg on the Mn/TiO_2 catalyst for selective catalytic reduction of NO_x with NH_3: a mechanistic study. J Phys Chem C. 2017;121:7881–91.
31. Liu J, Guo R, Sun X, et al. Simultaneous NO reduction and Hg^0 oxidation over Sb modified Mn/TiO_2 catalyst. Mater Chem Phys. 2019;232:88–98.
32. Sun X, Guo R, Liu J, et al. The enhanced SCR performance of Mn/TiO_2 catalyst by Mo modification: identification of the promotion mechanism. Int J Hydrogen Energy. 2018;43:16038–48.
33. Liu S, Guo R, Sun P, et al. The enhancement of Zn resistance of Mn/TiO_2 catalyst for NH_3-SCR reaction by the modification with $Al_2(SO_4)_3$. J Taiwan Inst Chem Eng. 2017;78:370–7.
34. Chen Q, Guo R, Wang Q, et al. The promotion effect of Co doping on the K resistance of Mn/TiO_2 catalyst for NH_3-SCR of NO. J Taiwan Inst Chem Eng. 2016;64:116–23.
35. Yang G, Zhao H, Luo X, et al. Promotion effect and mechanism of the addition of Mo on the enhanced low temperature SCR of NO_x by NH_3 over MnO_x/γ-Al_2O_3 catalysts. Appl Catal B. 2019;245:743–52.
36. Yang Q, Ren Y, Zhao L, et al. Development of catalyst for industrial flue gas denitration at low temperature. Appl Mech Mater. 2014;535:683–7.

37. Jin R, Liu Y, Wu Z, et al. Low-temperature selective catalytic reduction of NO with NH_3 over Mn-Ce oxides supported on TiO_2 and Al_2O_3: a comparative study. Chemosphere. 2010;78:1160–6.
38. Huang X, Zhang S, Chen H, et al. Selective catalytic reduction of NO with NH_3 over V_2O_5 supported on TiO_2 and Al_2O_3: a comparative study. J Mol Struct. 2015;1098:289–97.
39. Li G, Mao D, Chao M, et al. Low-temperature NH_3-SCR of NO over $MnCeO/TiO_2$ catalyst: enhanced activity and SO_2 tolerance by modifying TiO_2 with Al_2O_3. J Rare Earths. 2021;39:805–16.
40. Li RH, Zhou S, Li CL, et al. A study on SCR catalyst support. Adv Mater Res. 2013;726–731:17–20.
41. Pan X, Bao X. Reactions over catalysts confined in carbon nanotubes. Chem Commun. 2008;6271–81.
42. Youn J, Kim M, Lee S, et al. The influence of CNTs addition on Mn-Ce/TiO_2 catalyst for low-temperature NH_3-SCR of NO. Catal Commun. 2021;152: 106282.
43. Wang X, Zheng Y. Lin J Highly dispersed Mn–Ce mixed oxides supported on carbon nanotubes for low-temperature NO reduction with NH_3. Catal Commun. 2013;37:96–9.
44. Yang J, Ren S, Zhang T, et al. Iron doped effects on active sites formation over activated carbon supported Mn–Ce oxide catalysts for low-temperature SCR of NO. Chem Eng J. 2020;379: 122398.
45. Shen B, Chen J, Yue S, et al. A comparative study of modified cotton biochar and activated carbon based catalysts in low temperature SCR. Fuel. 2015;156:47–53.
46. Shu Y, Zhang F, Wang H. Manganese–cerium mixed oxides supported on rice husk based activated carbon with high sulfur tolerance for low-temperature selective catalytic reduction of nitrogen oxides with ammonia. RSC Adv. 2019;9:23964–72.
47. Jiang LJ, Liu QC, Zhao Q, et al. Promotional effect of Ce on the SCR of NO with NH_3 at low temperature over MnO_x supported by nitric acid-modified activated carbon. Res Chem Intermed. 2017;44:1729–44.

Chapter 6
Ceria-Based SCR Catalysts

6.1 Introduction

In the process of industrial production, nitrogen oxides (NO_x) emission raises serious environmental issues and severely endangers human health. In order to cushion these negative effects, selective catalytic reduction (SCR) with NH_3 has been widely applied for NO_x removal. However, the commercial SCR catalyst V_2O_5-WO_3(MoO_3)/TiO_2 has some serious limitations, for instance, the excessively higher operation temperature (> 300 °C). Therefore, to cater the demand for the low-temperature (L–T) SCR with NH_3, CeO_2 has been extensively investigated. Owing to CeO_2 has the advantages of satisfactory SO_2-tolerance, redox properties and oxygen storage capacity, which has the potential to be a promising low-temperature SCR catalyst.

With the in-depth research on ceria-based oxides, ceria is commonly used as promoter or support in NH_3 SCR catalysts due to its good structural properties and strong ability to interact with other components. Although pure CeO_2 generally exhibits poor NH_3-SCR activity, our previous studies have found that factors such as precursor, calcination temperature, preparation method and morphology have a significant impact on the catalytic performance of pure CeO_2 for the selective reduction of NO with NH_3. Furthermore, according to the state of cerium oxide in catalysts, it can be divided into three types: surface modification, bulk doping, pure support or active component. However, a comprehensive summary of ceria-based oxides catalysts with different compositions, supports, and morphologies is still insufficient.

In this section, we concentrate on ceria-based catalysts with an attempt to supply an update on recent developments. This chapter introduces catalysts from the following three types: single ceria-based catalysts, composite ceria-based catalysts, and Supported ceria-based catalysts, respectively. To clarify the various roles of each component in ceria-based catalysts, this review elaborated on in detail of a new

W. Pan and R. Guo, *Low-Temperature Selective Catalytic Reduction Catalysts*,
https://doi.org/10.1007/978-981-97-9298-6_6

perspective to understand the performance of the catalyst. Furthermore, the development trend of low-temperature NH_3-SCR catalysts were discussed, to advance the industrial applications of ceria-based catalyst.

6.2 Single Ceria-Based Catalysts

Generally, as rare earth elements exist with unfilled 4f orbitals and with lanthanide contraction, they have been widely applied as active components or catalyst supports in the catalytic combustion of fossil fuels. Among them, as one of the most abundant elements, cerium plays an exceedingly significant role in the field of low-temperature NH_3-SCR due to excellent oxygen storage capacity and environmental friendliness. Normally, the stoichiometric CeO_2 remained a face-centered cubic structure [1]. In this structure, per unit cell contained four cerium and eight oxygen atoms [2]. When the stoichiometric CeO_2 is reduced partially, the stoichiometric CeO_2 is converted into the nonstoichiometric CeO_2 promptly, evoking the formation of oxygen vacancies. Since CeO_2 has a unique structure, it exhibits a peculiar surface acid–base, excellent oxygen storage capacity, and outstanding redox properties. Based on the distinctive surface properties, CeO_2 seems to be a promising active component for the SCR of NO with NH_3. Therefore, researchers have explored a range of ceria-based catalyst. Along with the unceasingly thorough research, it has been demonstrated that the catalytic performance of a single cerium catalyst is connected with precursor, calcination temperature, preparation method, morphology and exposed crystalline surfaces.

6.2.1 Effect of Precursor and Calcination Temperature on Catalytic Performance

In our previous research, we found that the de-NOx performance of ceria-based catalyst was closely correlated with the cerium precursor and calcination treatment. According to a previous report, the catalytic activity of pure CeO_2 catalyst was not excellent, and the NO conversion was below 40% consistently [3]. After surface acidification, the active oxygen species on the pure CeO_2 catalyst surface increased and NH_3 chemisorption enhanced, both of which were conducive to NH_3 activation. Pan et al. [4] prepared pure CeO_2 catalysts from different cerium precursors (cerium nitrate and cerium acetate) by using the citric acid method and their activities for selective catalytic reduction of NO with NH_3 were investigated in the presence of O_2. The results showed that the catalysts prepared from cerium nitrate had better selective catalytic reduction performance and SO_2 resistance. As could be seen from the BET, XRD, TPR, TPD and XPS results, the good performance of the catalysts prepared from cerium nitrate should be resulted from large surface area, good redox ability, high NH_3 adsorption capacity and enrichment of Ce^{3+} on the surface accompanied by an increase of active oxygen content. Calcination temperature is another element with a momentous influence on the SCR performance of pure CeO_2. Guo et al. [5] investigated pure CeO_2 as a catalyst for selective catalytic reduction of NO by NH_3. The ceria catalysts were prepared by citric acid method with different calcination temperature. When the calcination temperature of CeO_2 was 350 °C, the pure CeO_2 catalyst manifested the optimal catalytic activity (Fig. 6.1). The SCR performance of CeO_2 catalysts decreases with increasing calcination temperature. In addition, when reaction temperature is below 250 °C, NO conversion increases with increasing reaction temperature; but when reaction temperature exceeds 250 °C, it would decrease with increasing reaction temperature, which may be due to an increase of surface energy below 250 °C and the oxidation of NH_3 to NO at high temperature. Experimental results showed that the selective catalytic reduction activity of ceria catalyst decreases with increasing calcination temperature within the reaction temperature range of 150–350 °C. As can be seen from the characterization results of BET, XRD, temperature programmed reduction, total programmed desorption and XPS, the best selective catalytic reduction performance of ceria catalyst calcinated at 350 °C may be resulted from its large surface area, good redox ability, strong NH_3 adsorption capacity and high concentration of active oxygen species on the surface of this catalyst. Therefore, the performance of ceria catalysts can be improved to some extent by choosing appropriate precursor and calcination temperature.

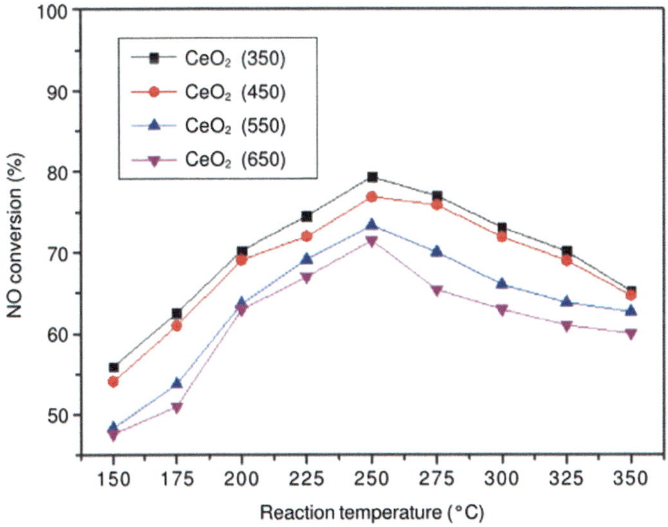

Fig. 6.1 NO conversion as a function of reaction temperature for different catalysts. Reaction conditions: [NO] = [NH_3] = 600 ppm, [O_2] = 5%, balance Ar, total flow rate = 1 L/min, GHSV = 28,000 h^{-1} [5]

6.2.2 *Effect of Preparation Method on Catalytic Performance*

Generally speaking, pure CeO_2 catalyst has two main drawbacks: lower specific surface area and higher deactivation due to thermal sintering, respectively. Appropriate preparation methods can partially improve the above defects. Sutthisripok et al. [6] synthesized pure CeO_2 catalysts employing three different methods: cerium oxide-based catalysts fabricated by the surfactant assisted method (CeO_2(SF)), cerium oxide-based catalysts prepared by the templating pathway (CeO_2(TP)), and cerium oxide-based catalysts constructed by precipitation method (CeO_2(PP)). It was evident that CeO_2 (SF) exhibited the largest specific surface area. Accordingly, an appropriate preparation method is propitious to augment a specific surface area, thereby facilitating the SCR efficiency. Yang et al. [7] investigated that the effect of HF treatment on the performance of ceria for selective catalytic reduction of NO with NH_3. The experimental results showed that HF treatment could greatly enhance the SCR activity of CeO_2 catalyst within the temperature range of 100–400 °C. From the results of characterizations, it could be found that the treatment of CeO_2 catalyst by HF would lead to lower crystallinity, better reducibility, stronger NH_3 adsorption ability and more surface adsorbed oxygen. As can be seen from the results of in situ DRIFT study, the NH_3-SCR reaction over CeO_2 and the treatment of CeO_2 catalyst by HF mainly follows the Eley–Rideal mechanism. Moreover, Sun et al. [8] prepared fresh CeO_2 and the NH_3-treated CeO_2 sample (CeO_2–N). The corresponding samples after the activity test from 100 to 400 °C were named as CeO_2-used-1 and CeO_2–N-used-1. The prepared samples after the stability test were labeled as CeO_2-used-2 and CeO_2–N-used-2 respectively. The SCR catalytic abilities of CeO_2 and CeO_2–N ranging from 100 to 400 °C are depicted in Fig. 6.2. In the preset temperature range, pure CeO_2 exhibits bad SCR activity of no more than 55%. It is remarkable that NH_3 treatment could effectively enhance its SCR performance. The activity of CeO_2–N is over 80% in 195–336 °C. Furthermore, the activities of the two samples after the activity test (CeO_2-used-1 and CeO_2–N-used-1) were also evaluated. Apparently, the NO_x conversion of CeO_2-used-1 decreases obviously compared to CeO_2. In contrast, CeO_2–N-used-1 exhibits a slight activity decrease in the whole reaction temperature range. Additionally, NH_3 treatment also helps to improve the N_2 selectivities of CeO_2 and CeO_2-used-1, as shown Fig. 6.2b. It was found that NH_3 treatment was an effective way to improve the SCR activity of CeO_2. Characterization results proved that NH_3 treatment of CeO_2 could decrease its crystallinity and increase its reducibility, accompanied with the generation of more surface adsorbed oxygen species and more Ce^{3+}. The qualities mentioned above contributed to the enhancement of the SCR activity and the stability of CeO_2–N. Furthermore, the results of DRIFT study showed that the SCR reaction over CeO_2–N was mainly accelerated via the L–H route.

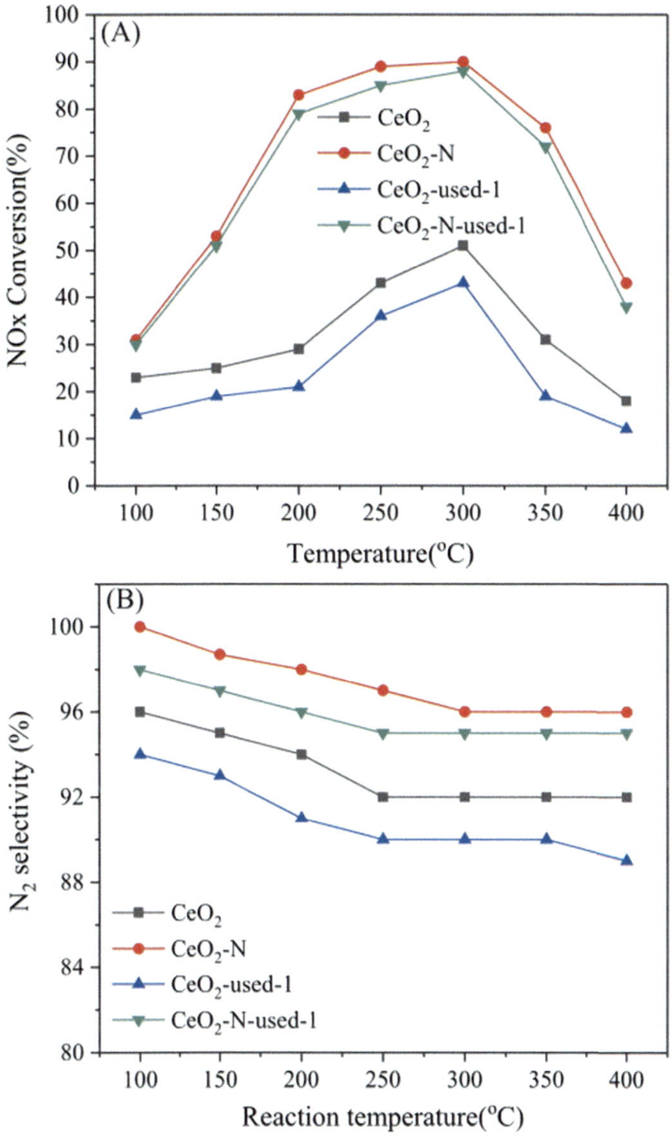

Fig. 6.2 **a** NO_x conversions **b** N_2 selectivities over CeO_2, CeO_2–N, CeO_2-used-1 and CeO_2-N-used-1 catalysts. Reaction conditions: [NO] = [NH_3] = 600 ppm, [O_2] = 5%, balance Ar, GHSV = 108,000 h^{-1} [8]

6.2.3 *Effect of Morphology and Exposed Crystalline Surfaces on Catalytic Performance*

Furthermore, the morphology and exposed crystalline surfaces of ceria-based catalyst also has a significant impact on catalytic performance. The Ce-based catalysts obtained by different synthetic routes could form various shapes, including nano-cubes, nano-polyhedrons, nano-rods. To investigate the effect of CeO_2 nanomaterial morphology on its performance for NO catalytic oxidation. Wang et al. [9] prepared three kinds of CeO_2 nanomaterials including CeO_2 nanorods, nanospheres and nanoparticles by hydrothermal method and used for catalytic oxidation of NO at low temperature (Fig. 6.3). The experimental results show that CeO_2 nanorods are of the best catalytic performance. Characterization techniques including TEM, XRD, H_2-TPR, NO-TPD and XPS were used to determine the relationship between the morphology of CeO_2 nanomaterial and its catalytic performance. TEM images show that CeO_2 nanorods predominantly exposed (110) and (100) planes, while CeO_2 nanospheres and CeO_2 nanoparticles predominantly exposed (111) plane. The excellent catalytic performance of CeO_2 nanorods could be ascribed to the low crystallinity, high reducibility, strong NO adsorption ability and the presence of more surface chemisorbed oxygen. With the in-depth research on ceria-based oxides, it can be inferred that the morphology of ceria-based catalyst has significant effects on its stability and catalytic activity. For example, Guo et al. [10] fabricated a CeO_x@MnO_x core–shell structure catalyst through depositing Mn precursor onto CeO_2 nanoparticles, which displayed superior intrinsic catalytic performance than the $CeMnO_x$ catalyst without core–shell structure. Likewise, Liu and co-authors synthesized a MnO_x–CeO_2 catalyst with shell-in-shell microsphere structure [11]. Compared with a MnO_x–CeO_2 without this unique structure, the former was endowed with larger specific surface area and higher oxygen mobility through the method of morphology control. The conventional ceria-based supported catalysts usually have drawbacks of inferior dispersibility and easier aggregation. In contrast, the ceria-based catalyst with core–shell structure for SCR with NH_3 have the following merits: (1) The core–shell structure has a markedly impact on the NH_3-SCR performance, since the characteristic morphology and crystal-plane effects can exert the interaction between core and shell to improve the redox capacity and surface acidity of catalyst. (2) The shell acts as a protective layer, which can ameliorate the thermal stability and SO_2 resistance of catalyst to a certain extent. (3) By protecting the core material, the migration and aggregation of nanoparticles can be inhibited. Therefore, the SCR performance of ceria-based catalysts can also be improved by the method of morphology control.

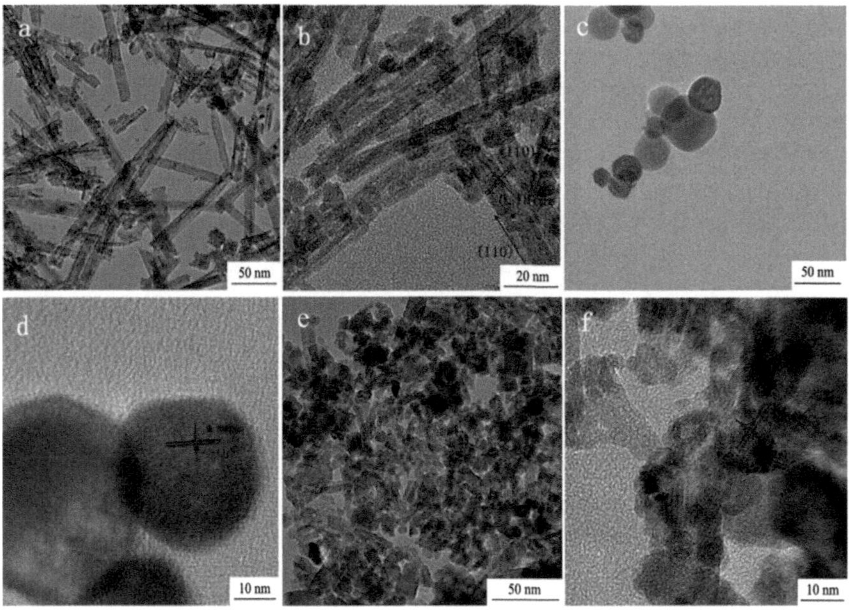

Fig. 6.3 TEM and high-resolution TEM images. **a**, **d** CeO_2 nanorods; **b**, **e** CeO_2 nanospheres; **c**, **f** CeO_2 nanoparticles [9]

6.3 Composite Ceria-Based Catalysts

Compared with ceria-based supported catalysts, cerium-based mixed oxide catalysts are dominant in NH_3-SCR. Actually, bulk modification of CeO_2 is a common means to improve the catalytic performance and thermal stability. The properties of CeO_2 can be modulated by the addition of appropriate metal oxides to obtain composite ceria-based catalysts. In composite oxide catalysts, one metal element can modify the catalytic properties of another, which results from both electronic and structural influences. After bulk modification of CeO_2, they are increasingly attractive due to their high activity. Other metal oxides are doped to modify the catalytic performance of pure CeO_2 catalysts.

6.3.1 Mn–Ce Composite Oxide System

As mentioned above, pure CeO_2 catalysts are inapplicable to de-NOx in low-temperature SCR. Currently, by doping other transition metals, Ce-based catalysts can be modified for low-temperature de-NO_x reaction. The impact of different dopants on CeO_2 catalysts for the SCR de-NO_x with NH_3 has been studied. MnO_x has been considered an ideal catalyst for low-temperature SCR of NO with NH_3.

Therefore, MnO_x–CeO_2 is well known for its excellent low-temperature NO reduction performance and has attracted great interest. Qi et al. [12] investigated that the factors which influence the activity of MnO_x–CeO_2 catalysts, including the molar ratio of Ce and Mn, calcination temperature, and preparation methods, were determined and compared. The result indicated that the low-temperature SCR activity of MnO_x–CeO_2 catalysts was affected by the different molar ratio of Ce and Mn significantly. Especially when the molar ratio of Mn/(Mn + Ce) was 0.3, the MnO_x–CeO_2 catalyst maintained nearly 100% NO conversion at above 120 °C. The MnO_x–CeO_2 catalysts were calcined at 400, 500, 650 and 750 °C, and their SCR activities were measured. When the calcination temperature is 650 °C, the result indicated that the catalyst had the highest catalytic performance. Besides, the catalytic performances of MnO_x–CeO_2 catalysts constructed by different methods were analyzed and evaluated. The MnO_x–CeO_2 catalyst prepared by the citric acid method revealed the highest NO conversion. However, the specific surface area of MnO_x–CeO_2 catalyst synthesized by the citric acid method is not the largest among them. Consequently, the surface area is a considerable incentive for low-temperature SCR activity, but not the only one. In the following research, the results of XRD, Fourier transform infrared spectroscopy (FTIR), and X-ray photoelectron spectroscopy (XPS) revealed that there were strong interactions between Ce and Mn elements. Hence, the exceptional SCR activity could be attributed to the highly dispersed Mn species. Furthermore, MnO_x–CeO_2 catalysts doped with a third metal could further improve the SCR efficiency. Casapu et al. [13] investigated which different dopants, such as Nb, Fe, W, and Zr oxides, were applied in the modification of MnO_x–CeO_2 catalysts. In addition, among different metal oxides doped to MnO_x–CeO_2, only MnO_x–CeO_2 catalysts doped with Nb oxides markedly increased SCR activity and almost 80% NO conversion at above 120 °C.

6.3.2 Ce–Cu Composite Oxide System

Recently, Cu was found to be effective in enhancing the low temperature SCR activity of cerium-based catalyst. And the state of Cu species had great impact on the low temperature catalytic activity of the catalyst. It is well known the preparation methods and preparation process parameters have great impact on the activity of the catalyst. As an important process parameter, calcination temperature would affect the particle size, surface concentration, and oxidation state of active components of a SCR catalyst. Guo et al. [14] prepared CeO_2–CuO catalyst by citric acid method to investigate for selective catalytic reduction of NO with NH_3. The activity of the CeO_2 catalyst was enhanced by the addition of Cu. The characterization results showed that the decreased CeO_2 crystallinity, stronger acidity, better oxygen storage capacity and more appearance of Ce^{3+} may be the main reasons for its good low temperature SCR performance. The state of Cu species has great impact on the SCR activity of CeO_2–CuO catalyst. Cu^{2+} can improve the low temperature activity, while CuO would decrease the high temperature activity. Pan et al. [15] prepared CeO_2–CuO

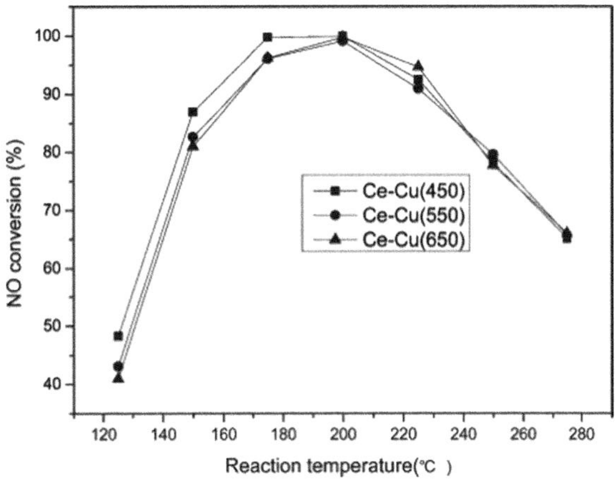

Fig. 6.4 NO conversion over the CeO_2–CuO catalysts. Reaction conditions: total flow rate = 1L/min, $[NH_3] = [NO] =$ 600 ppm, $[O_2] = 5\%$, Ar balance, GHSV = 28,000 h^{-1} [15]

catalysts by citric acid method at different calcination temperatures to investigate the influence of calcination temperature on the catalytic activity of CeO_2–CuO catalyst for the selective catalytic reduction of NO with NH_3. As shown in Fig. 6.4, Ce–Cu (450 °C) is of the best low temperature selective catalytic reduction (SCR) activity among the three catalysts, which may be resulted from its large surface area, good redox ability, strong NH_3 adsorption capacity, and high concentration of Cu^{2+} and active oxygen species on the surface.

6.3.3 Ce–Ti Composite Oxide System

Due to the excellent redox performance of cerium oxide and good acidity of titanium dioxide, it is reasonable to assume the binding of cerium oxide and titanium dioxide in NH_3-SCR. Compared with VO_x/titanium dioxide catalyst, CeO_2/TiO_2 has the advantage of environmental friendliness. In addition, cerium oxide can obtain higher surface area, which indicates that cerium oxide can be used as a carrier to form a new structure of TiO_2/CeO_2 in addition to being used as a surface dispersing component. In order to further improve the low temperature SCR performance of CeO_2 catalysts, one or more metal (Ho, Sm and Zr) ions can be doped to improve the performance. At the same time, these metal ions not only produce a large number of oxygen vacancies, which can improve the redox performance, but also promote the surface acidity of the catalysts through multicomponent co-catalysis. Duan et al. [16] discovered that the addition of Ho on CeTi catalyst could remarkably strengthen its K^+ tolerance and low-temperature SCR performance. The CeHoTi possesses the highest BET surface area and both Ce and Ho species were highly scattered on the catalyst surfaces. Moreover, the higher concentration of Ce^{3+} and more chemisorbed oxygen was beneficial to the improved redox capacity and the enhancement of the "fast SCR" process. The addition of Ho generates more acid sites, which could enhance the adsorption of NH_3 species. Moreover, in situ DRIFT analysis indicated that additional ad-NH_3 species and ad-NO_x species with higher reactivity could be discovered on the surface of CeHoTi, which was also crucial for the improved K resistance. And the NH_3-SCR reactions over CeTi and CeHoTi primarily taked place through the Langmuir–Hinshelwood (L–H) pathway (Fig. 6.5). Sun et al. [17] introduced Sm into Ce/TiO_2 catalyst to enhance its low temperature SCR performance. The experimental results demonstrated that CeSm/TiO_2-0.3 catalyst presented an excellent low-temperature SCR activity. From the characterization results, it could be found that the modification of Ce/TiO_2 catalyst by Sm could generate more Ce^{3+} and chemisorbed oxygen species, therefore, promoting its reducibility and the oxidation of NO to NO_2, accompanied by the enhanced NH_3 adsorption. As a result, the NH_3-SCR reaction over CeSm/TiO_2-0.3 was greatly facilitated in the low temperature range (< 200 °C). Liu et al. [18] reached a similar conclusion by doping Zr. The BET, XRD, Raman, XPS, H_2-TPR, NH_3-TPD, O_2-TPD and NO_x-TPD were adopted for exploring the mechanism of the promoted anti K-poisoning property of Ce/TiO_2 catalyst. The obtained results demonstrated that Ce/TiO_2 with the addition of Zr exhibited larger specific surface area, higher dispersion of Ce species, higher redox property and more acid sites. By virtue of the in situ DRIFT analysis, the SCR reaction over Ce/$TiZrO_x$ catalyst was accelerated via Langmuir–Hinshelwood (L–H) mechanism. Besides, the addition of non-metallic ions (Bi and P) in our research also improved the performance of Ce/TiO_2 catalysts [19, 20].

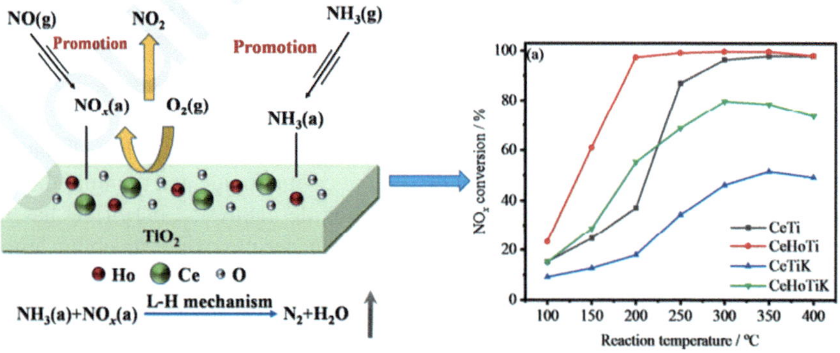

Fig. 6.5 The addition of Ho could remarkably strengthen the K tolerance of CeTi catalyst, and its low-temperature SCR performance has also been improved [16]

6.4 Supported Ceria-Based Catalysts

Generally, the function of CeO_2 in supported catalysts can be roughly classified into three main types: as an active ingredient, as a cocatalyst, and CeO_2 as a support. To elucidate the contact among composition-structure–activity of the supported catalysts, substantial investigations have been implemented at the levels of molecular and atomic. Afterward, researchers proposed the supports of supported catalysts normally contains two types: rigid supports (Al_2O_3, SiO_2, etc.) and active supports (TiO_2, etc.). Furthermore, the synergistic reaction between active supports and surface loading components is extremely dramatic, while rigid supports are not. Fortunately, CeO_2 is also an active support, exhibiting respectable surface-bound defects, satisfactory oxygen storage capacity, and fabulous redox ability. In this section, following the sequence from CeO_2 as the support then to as the surface loading component, the function of CeO_2 is described systematically over the supported catalysts.

6.4.1 *CeO_2 as the Support*

Since active components are remarkably dispersed over the surface of pure CeO_2, researchers employed pure CeO_2 as the support of catalysts. Additionally, researchers are trying to illuminate the mechanism of performance enhancements by adopting pure CeO_2 as the support. Loading active components on the surface of pure CeO_2 not only enhanced the dispersion of active species but also heightened the synergism in the interfacial region. Meanwhile, as the support, pure CeO_2 ameliorated redox properties of surface dispersed active species [21]. However, although the pure CeO_2 support has massive advantages, the weaknesses should not be neglected, for example, smaller specific surface area and easier sintering. Furthermore, to acquire maximum activity, it is extraordinarily essential that the pure CeO_2 support should be in contact with active components. In this context, the utilization of pure CeO_2 as a support seems to be unsatisfactory, therefore, to a certain extent, restricting its generalization.

6.4.2 *CeO_2 as the Surface Loading Component*

Supports, supplying tremendous surface area that is conducive to the dispersion of active phases and, in the meanwhile, fabricating massive synergistic sites for the activation of reactants, ameliorate the crystalline particles of catalysts. Thus far, carriers of various materials, including TiO_2, Al_2O_3, active carbon (AC), and zeolite, etc., have been adopted to ceria-based low-temperature NH_3-SCR. The merits and demerits of various carriers have been investigated by way of contrast, to screen out the most satisfactory carrier. Meanwhile, the carrier often affects the valence state of metal, which further affects the redox performance. Therefore, a targeted carrier of catalyst with large specific surface area and enriched surface chemical site is necessary.

(1) **TiO_2 as the support**

TiO_2 as a support is nonpoisonous, is readily available, has relatively fabulous thermal stability, and was employed as the most widespread for the low-temperature NH_3-SCR. Li et al. elucidated by in situ FTIR, XPS, X-ray absorption fine-structure (XAFS), and field-emission TEM (FE-TEM) results that due to the synergistic effect between Ce and Ti at the atomic scale, for the first time, the Ce–O–Ti short-range order species was identified as the active site [22]. Wang et al. found that the presence of active energy (001) facets on the TiO_2 support facilitates thermal stability, and at the same time, the Ti^{3+} on the TiO_2 surface accelerates the SCR process, thus modulating the catalytic performance of CeO_2 catalysts [23]. Unexpectedly, the catalytic performance is unsatisfactory at low temperature. In order to amend the low-temperature efficiency of CeO_2 catalysts, introducing other elements modifications is the most common method.

Sun and coauthors investigated that the addition of Sm, facilitating the adsorption of NH_3 species and the oxidation of NO, could heighten the low-temperature SCR activity [17]. Characterization results indicated that the introduction of Sm could promote the formation of Ce^{3+} and chemisorbed oxygen species on catalyst surface, which was favorable to the generation of NO_2. Moreover, the doping of Sm could also promote the adsorption of NH_3 and NO_x species on Ce/TiO_2 catalyst. The results of in situ DRIFT study revealed that the NH_3-SCR reaction over CeSm/TiO_2 catalyst was mainly controlled by the Langmuir–Hinshelwood mechanism in the low temperature range (< 200 °C). Thereafter, the promoted NO oxidation and the enhanced NH_3 adsorption should play a major role for the excellent low-temperature SCR performance of CeSm/TiO_2 catalyst. Guo et al. educed a similar conclusion through doping Bi into CeO_2/TiO_2 catalysts [19]. In general, the introduction of appropriate elements increases the specific surface area, improves the surface acidity, and is conducive to improving the performance of low-temperature SCR.

(2) **Al_2O_3 as the support**

Aluminum oxide (Al_2O_3) a s a satisfactory catalyst support and adsorbent has been employed in the field of catalysis vastly. With Al_2O_3, including several transitional phases among them, γ-Al_2O_3 (activated alumina) exhibited enormous surface area, fabulous porosity, outstanding surface acidity, and thermodynamic stability. γ-Al_2O_3 demonstrated outstanding adsorption performance, which was attributed to immense pore volume, additionally minimizing the influences of diffusion and transport [24].

Aluminum oxide, a common carrier like TiO_2, has been extensively studied for the low-temperature SCR of NO_x. With further research on CeO_2/Al_2O_3 catalyst, extensive results manifested that preparation method dramatically affected the catalytic performance of CeO_2/Al_2O_3 catalyst. Guo et al. contrasted the three different preparation methods: sol–gel, impregnation, and coprecipitation [25]. As illustrated in Fig. 6.6, the experimental results proved that CeO_2/Al_2O_3 catalyst constructed by

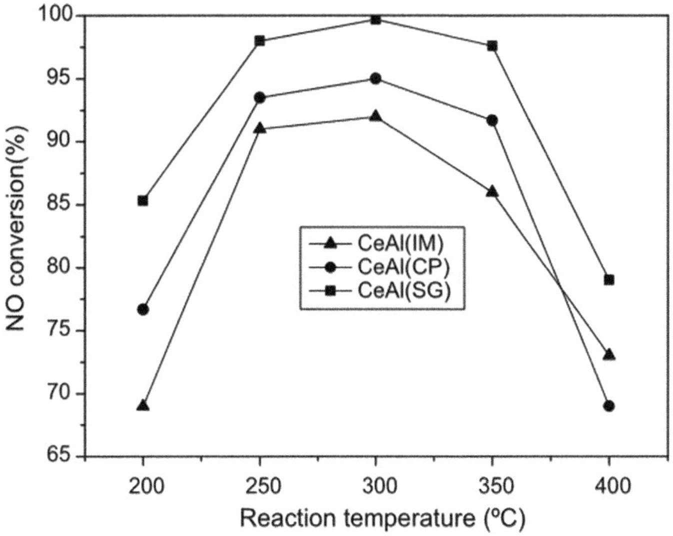

Fig. 6.6 Catalytic activity of CeO_2/Al_2O_3 catalyst samples. Reaction conditions: [NO] = [NH_3] = 600 ppm, [O_2] = 5%; balance Ar, total flow rate = 1 L/min, GHSV = 25,000 h^{-1}

the sol–gel method exhibits satisfactory SCR activity in the temperature range of 200–400 °C. As the results of BET, XRD, NH_3-TPD and H_2-TPR characterization, strong interaction, large surface area, highly dispersed nano-crystalline ceria, high NH_3 adsorption capacity and good redox ability should be the main reasons for the excellent performance of the CeO_2/Al_2O_3 catalyst prepared by the single step sol–gel method. Therefore, it indicates that the catalyst preparation method has great impact on the catalytic performance of CeO_2/Al_2O_3 catalyst.

(3) **Carbon materials as the support**

Carbon materials, such as activated carbon (AC), activated carbon fibers (ACFs), and carbon nanotubes (CNTs), are provided with the characteristics of massive surface area, exceptional porous structure, and colossal pore volume. Commonly, activated carbon was modified in the pretreatment process to amend the low-temperature SCR performance. Wang et al. synthesized excellent activated carbon support by adopting phenolic resin as a binder for the immobilization of MnO_x–CeO_2 catalyst [26]. The activated carbon support, exhibiting extraordinary honeycomb structures, was propitious to the dispersion of active species, markedly.

From the angle of catalysis, compared with the traditional carbon materials (AC, ACFs), the CNTs was a promising alternative to AC or ACFs. Substantial research has elucidated the utilization of CNTs as the support could boost the activity and selectivity and the consequently be employed as the support material for low-temperature SCR. Nevertheless, on account of CNTs agglomeration readily, it is enormously crucial to ameliorate the dispersion of CNTs. Currently, the predominant avenues to improve the blemish are divided into the following several points: ball milling, molecular mixing, in situ, and solution coating [27]. Unfortunately, several limitations emerged by the above method. For instance, CNTs supports fabricated through the in situ method might generate impurities.

The fabulous properties of graphene, the support for low temperature SCR, were attributed to the oxygen-containing functional groups on the surface of catalysts. Through oxidation treatment, oxygen-containing functional groups (such as carboxyl, carbonyl, and hydroxyl) can be introduced to chemically modify the outer surface of graphite to facilitate the NH_3-SCR activity at low-temperature. Zhou et al. emphasized that by applying graphene oxide sheets (GO) as a support, the CeO_2/GO catalysts exhibited better catalytic performance in the temperature range of 175–325 °C [28]. The morphology, structure and composition of 10% CeO_2/GO catalysts were investigated by high angle annular darkfield-scanning transmission electron microscopy (HAADF-STEM). Moreover, the distribution of C, O and Ce elements was analyzed and confirmed by EDX. Figure 6.7a–e shows the HAADF-STEM images of 10% CeO_2/GO and corresponding EDX elemental maps of C, O and Ce. The EDX maps of 10% CeO_2/GO catalyst show that C, O and Ce elements coexist uniformly, indicating the close contact between CeO_2 and GO. 2D sheet morphology of typical GO can be observed by AFM images of 10% CeO_2/GO as shown in Fig. 6.7f, The HRTEM image (Fig. 6.7g) clearly reveals the lattice fringes with the lattice spacing of 0.26 nm and 0.31 nm, which correspond well to CeO_2 (220) and (111) planes. According to characterization results, it can be seen that CeO_2 can be successfully doped on the GO by simple impregnation method. In addition, the two-dimensional sheet morphology can provide large specific surface areas to adsorb the gaseous reactants. The enhanced performance of 10% CeO_2/GO catalyst should be attributed to the interaction between CeO_2 and GO, which greatly improves the reducibility of catalyst. The excellent SCR activity of the 10% CeO_2/GO catalyst is due to the abundant Ce^{3+}, chemisorbed oxygen and Lewis acid sites.

(4) **Zeolites as the support**

Zeolites, crystalline aluminosilicates, manifested excellent ion exchange properties and abundant surface acidity owing to the replacement of Si ions with an Al ion. Contemporarily, miscellaneous structures of zeolite have been authenticated, which could be discriminated through the maximum aperture of the frame structure opening, for instance, weeny (8-membered ring), middling (10-membered ring), and enormous and king-sized (12-membered ring or above), respectively [29]. As a carrier, it is noteworthy that zeolites have superior adsorption performances and flexibility of complicated operating environments.

However, compared with conventional middling and enormous pore zeolites, the weeny pore zeolites have substantial distinguishing superiority: wider temperature window, outstanding thermal stability, and exceptionally brilliant SO_2 resistance performance. However, compared with conventional middling and enormous pore zeolites, the weeny pore zeolites have substantial distinguishing superiority: wider temperature window, outstanding thermal stability, and exceptionally brilliant SO_2 resistance performance. Recently, Yan et al. constructed an ultrasmall size nano-zeolite EMT precursor, which was employed to synthesize the Ce-$Si_5Al_2O_x$ catalyst

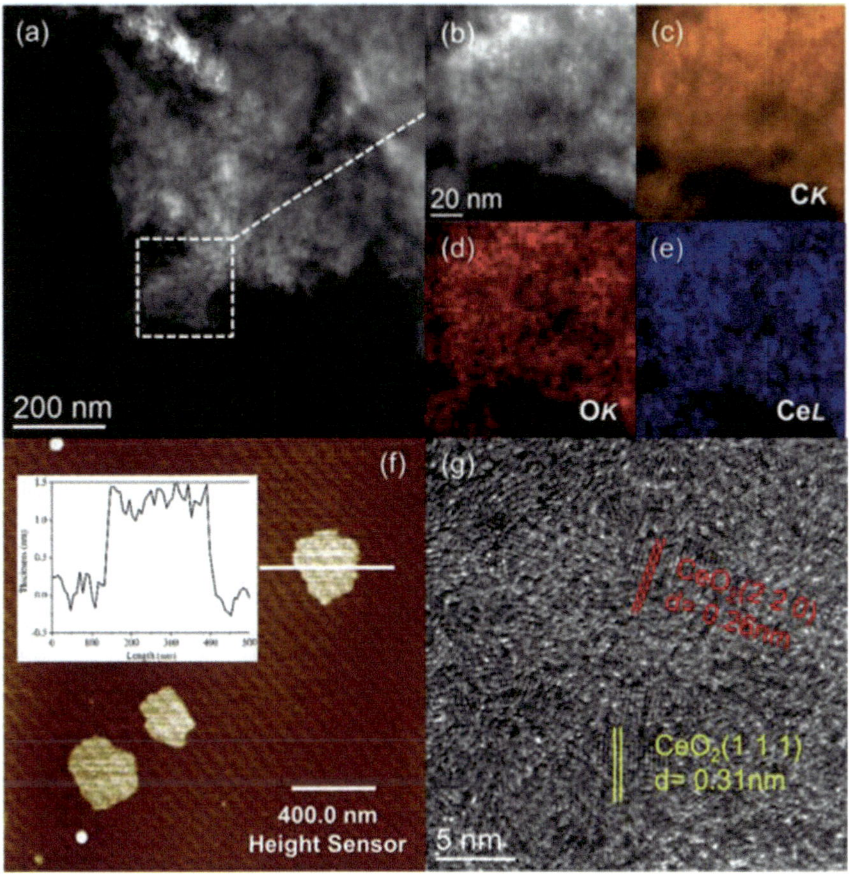

Fig. 6.7 **a** Low-magnification HAADF-STEM image; **b** high resolution HAADF-STEM image; **c–e** corresponding EDX elemental maps of C, O and Ce; **f** AFM image and height scanning line; **g** HRTEM image [28]

[30]. The results reflected that the Ce-$Si_5Al_2O_x$ catalyst, demonstrating the satisfactory resistance of K and P, was ascribed to highly ordered bonding between cerium and aluminosilicate. Currently, ion-exchange, impregnation, and one-pot hydrothermal methods have been applied for the synthesis of weeny pore zeolites. Guo et al. fabricated the Ce modified Cu-SSZ-13 catalysts via the wet impregnation method [31]. Based on the characterization results, it was found that the addition of Ce on Cu-SSZ-13 catalyst could improve its K tolerance. Relevant characterization results suggested that the modification with Ce could improve catalyst redox ability and form more surface adsorbed oxygen. Moreover, the adsorption capacity of NO and NH_3 over the catalyst was strengthened and the surface acidity was also increased, which was attributed to the synergetic effect that occurred on the zeolite

framework of Ce and Cu. Therefore, the catalyst modified with Ce had good resistance to K poisoning and low-temperature SCR performance. Ma et al. fabricated a Cu–Ce Co-doped SAPO-5/34 catalyst via the one-pot hydrothermal method, which exhibited a hybrid crystal structure [32]. The hybrid crystal structure was propitious to broaden the temperature window. Meanwhile, the introduction of Ce could facilitate the formation of acid sites and ameliorate the redox cycles of the catalyst. Weeny pore zeolites have demonstrated low temperature NOx reduction with stability at higher temperatures. SSZ-13 and SAPO-34 are by far the best support of catalysts suitable for a broader temperature window and also operate well below 350 °C with good hydrothermal stability. Optimum values of Si/Al ratio (SAR) play a crucial role and hence more studies on variation in SAR and adsorbed monodentate nitrates are required to get a fundamental understanding on their impact on low temperature SCR activity of Ce based catalysts with zeolites as the support.

References

1. Ma Y, Gao W, Zhang Z, et al. Regulating the surface of nanoceria and its applications in heterogeneous catalysis. Surf Sci Rep. 2018;73:1–36.
2. Ganduglia-Pirovano MV, Hofmann A, Sauer J. Oxygen vacancies in transition metal and rare earth oxides: current state of understanding and remaining challenges. Surf Sci Rep. 2007;62:219–70.
3. Gu T, Liu Y, Weng X, et al. The enhanced performance of ceria with surface sulfation for selective catalytic reduction of NO by NH_3. Catal Commun. 2010;12:310–3.
4. Pan W, Zhou Y, Guo R, et al. Effect of cerium precursor on the performance of pure CeO_2 catalysts for selective catalytic reduction of NO with NH_3. Asian J Chem. 2013;25:9079–82.
5. Guo R, Zhen W, Zhou Y, et al. Selective catalytic reduction of NO by NH_3 over ceria: effect of calcination temperature on the activity of catalysts. Asian J Chem. 2014;26:15416.
6. Sutthisripok W, Sattayanurak S, Sikong L. Effect of specific surface area on oxygen storage capacity (OSC) and methane steam reforming reactivity of CeO_2. J Porous Mater. 2008;15:519–25.
7. Yang N, Guo R, Tian Y, et al. The enhanced performance of ceria by HF treatment for selective catalytic reduction of NO with NH_3. Fuel. 2016;179:305–11.
8. Sun X, Guo R, Liu S, et al. The promoted performance of CeO_2 catalyst for NH_3-SCR reaction by NH_3 treatment. Appl Surf Sci. 2018;462:187–93.
9. Wang W, Guo R, Pan W, et al. Low temperature catalytic oxidation of NO over different-shaped CeO_2. J Rare Earths. 2018;36:588–93.
10. Guo R, Chen Q, Ding H, et al. Preparation and characterization of CeO_x@MnO_x core–shell structure catalyst for catalytic oxidation of NO. Catal Commun. 2015;69:165–9.
11. Liu C, Gao G, Shi J, et al. MnO_x–CeO_2 shell-in-shell microspheres for NH_3-SCR de-NO_x at low temperature. Catal Commun. 2016;86:36–40.
12. Qi G, Yang R. Performance and kinetics study for low-temperature SCR of NO with NH_3 over MnO_x–CeO_2 catalyst. J Catal. 2003;217:434–41.
13. Casapu M, Kröcher O, Elsener M. Screening of doped MnO_x–CeO_2 catalysts for low-temperature NO-SCR. Appl Catal B. 2009;88:413–19.
14. Guo R, Zhen W, Pan W, et al. Effect of Cu doping on the SCR activity of CeO_2 catalyst prepared by citric acid method. J Ind Eng Chem. 2014;20:1577–80.
15. Pan W, Zhou Y, Guo R, et al. Influence of calcination temperature on CeO_2–CuO catalyst for the selective catalytic reduction of NO with NH_3. Environ Prog Sustain Energy. 2014;33:385–9.

16. Duan C, Guo R, Liu Y, et al. Enhancement of potassium resistance of Ce–Ti oxide catalyst for NH_3-SCR reaction by modification with holmium. J Rare Earths. 2022;40:49–56.
17. Sun P, Guo R, Liu S, et al. Enhancement of the low-temperature activity of Ce/TiO_2 catalyst by Sm modification for selective catalytic reduction of NO_x with NH_3. Mol Catal. 2017;433:224–34.
18. Liu S, Guo R, Sun X, et al. Selective catalytic reduction of NO_x over Ce/$TiZrO_x$ catalyst: the promoted K resistance by $TiZrO_x$ support. Mol Catal. 2019;462:19–27.
19. Guo R, Li M, Sun P, et al. Mechanistic investigation of the promotion effect of Bi modification on the NH_3–SCR performance of Ce/TiO_2 catalyst. J Phys Chem C. 2017;121:27535–45.
20. Li M, Guo R, Hu C, et al. The enhanced resistance to K deactivation of Ce/TiO_2 catalyst for NH_3-SCR reaction by the modification with P. Appl Surf Sci. 2018;436:814–22.
21. Garcia PD, Juan A, Irigoyen B. Mn-doped CeO_2: DFT+ U study of a catalyst for oxidation reactions. J Phys Chem C. 2013;117:18063–73.
22. Li P, Xin Y, Li Q, et al. Ce–Ti amorphous oxides for selective catalytic reduction of NO with NH_3: confirmation of Ce–O–Ti active sites. Environ Sci Technol. 2012;46:9600–5.
23. Wang H, Cao S, Fang Z, et al. CeO_2 doped anatase TiO_2 with exposed (001) high energy facets and its performance in selective catalytic reduction of NO by NH_3. Appl Surf Sci. 2015;330:245–52.
24. Stanislaus A, Al-Dolama K, Absi-Halabi M. Preparation of a large pore alumina-based HDM catalyst by hydrothermal treatment and studies on pore enlargement mechanism. J Mol Catal A: Chem. 2002;181:33–9.
25. Guo R, Zhou Y, Pan W, et al. Effect of preparation methods on the performance of CeO_2/Al_2O_3 catalysts for selective catalytic reduction of NO with NH_3. J Ind Eng Chem. 2013;19:2022–5.
26. Wang Y, Ge C, Zhan L, et al. MnO_x–CeO_2/activated carbon honeycomb catalyst for selective catalytic reduction of NO with NH_3 at low temperatures. Ind Eng Chem Res. 2012;51:11667–73.
27. Zhou J, Guo R, Zhang X, et al. Cerium oxide-based catalysts for low-temperature selective catalytic reduction of NO_x with NH_3: a review. Energy Fuels. 2021;35:2981–98.
28. Zhou Z, Wang Y, Yuan Q, et al. Facile preparation of CeO_2 supported on graphene oxide sheets for NH_3-SCR: improvement of catalytic activity and SO_2 tolerance. ChemistrySelect. 2021;6:4859–65.
29. Mohan S, Dinesha P, Kumar S. NO_x reduction behaviour in copper zeolite catalysts for ammonia SCR systems: a review. Chem Eng J. 2020;384:123253.
30. Yan L, Ji Y, Wang P, et al. Alkali and phosphorus resistant zeolite-like catalysts for NO_x reduction by NH_3. Environ Sci Technol. 2020;54:9132–41.
31. Guo D, Guo R, Duan C, et al. The enhanced K resistance of Cu-SSZ-13 catalyst for NH_3-SCR reaction by the modification with Ce. Mol Catal. 2021;502:111392.
32. Zhou X, Chen Z, Guo Z, et al. One-pot hydrothermal synthesis of dual metal incorporated CuCe-SAPO-34 zeolite for enhancing ammonia selective catalytic reduction. J Hazard Mater. 2021;405:124177.

Chapter 7
Cu-Based and Fe-Based SCR Catalysts

7.1 Introduction

Copper-based catalysts are ideal active ingredients for NH_3-SCR catalysts due to their excellent redox ability and suitable acidity. At present, research on copper-based catalysts for denitrification reactions mainly focuses on copper-based molecular sieve catalysts as the main research object. Copper-based molecular sieve catalysts have good activity and N_2 selectivity in low-temperature environments, making them a promising medium to low-temperature denitrification catalyst. However, there is relatively little research on copper-based oxide catalysts, and more research is needed.

Iron-based catalysts have received widespread attention and application due to their environmentally friendly and low-cost characteristics, but due to the low-temperature activity of iron oxide in iron-based materials is insufficient, and it is necessary to dope other transition metals to enhance its low-temperature activity; Active substances such as iron oxide are prone to aggregation, and suitable materials are needed as carriers to improve the dispersibility of the active substances.

7.2 Cu-Based SCR Catalysts

7.2.1 *Copper Oxide-Based Catalyst*

In order to improve and enhance the stability, temperature window, selectivity, etc. of CuO catalysts. In recent years, researchers have introduced various metal oxides into CuO to improve the stability of pure CuO catalysts. For example, our group previously reported a CeO_2–CuO catalyst [1] prepared by citric acid method, which exhibited high denitrification efficiency in the low temperature range of 180–200 °C.

W. Pan and R. Guo, *Low-Temperature Selective Catalytic Reduction Catalysts*,
https://doi.org/10.1007/978-981-97-9298-6_7

Liu et al. [2] also reported that adding Cu to Ce–Ti catalysts can suppress high-temperature denitrification activity, but Cu can enhance their low-temperature denitrification activity. Wang et al. [3] successfully prepared porous CuO/Cu_2O heterostructures, as well as pure CuO and Cu_2O, using three different coordination types of Cu. Compared with pure CuO and Cu_2O, CuO/Cu_2O heterostructures exhibit good NH_3-SCR activity and N_2 selectivity in the low temperature range of 170–220 °C, even at high altitude speeds of 80,000 h^{-1}. In addition, the CuO/Cu_2O heterostructure exhibits excellent resistance to H_2O, SO_2, and alkali metal poisoning. Compared with pure CuO and Cu_2O, CuO/Cu_2O heterojunctions have larger pore volumes, higher specific surface area, stronger acidity, and more Lewis acidic sites.

Liu et al. [4] doped europium into Cu/TiO_2 catalysts: $CuEu/TiO_2$-0.15 exhibited a denitrification activity of 90% at 200–300 °C, with high N_2 selectivity and SO_2 resistance. From the characterization results, the $CuEu/TiO_2$-0.15 catalyst itself has a high specific surface area, and Cu and Eu are highly dispersed on the surface of the catalyst. In addition, the reaction between Eu and Cu also inhibits CuO crystallization, effectively promoting the reduction ability of the catalyst, while providing more chemically adsorbed oxygen and surface Cu^{2+}, which helps to generate NO_2 species and induce a "fast SCR" reaction.

7.2.2 Copper Based Molecular Sieve Catalyst

The application of Cu/SAPO-34 as a mainstream SCR catalyst in the aftertreatment system of automotive diesel engine exhaust has been widely studied. Cu/SAPO-34 and Cu/SSZ-13 catalysts have high N_2 selectivity and stability, but their hydrothermal stability is poor. [5] Wang et al. [6] found that the structure of Cu/SAPO-34 zeolite can be destroyed by water vapor at room temperature, leading to a significant decrease in NH_3-SCR catalytic activity. Therefore, in the industrial application of NH_3-SCR, it is necessary to simultaneously consider the hydrothermal stability of copper based molecular sieve catalysts at low and high temperatures. Peng et al. [7] found that after low-temperature hydrothermal treatment of Cu/SAPO-34 molecular sieve catalysts with different grain sizes, Cu/SAPO-34 with smaller particle size retained more active site for separating Cu^{2+} than Cu/SAPO-34 with larger particle size, thus showing higher NH_3-SCR activity. In addition, in SAPO-34 with aluminum phosphate structure, phosphorus atoms are one of the key factors for stabilizing the framework of SAPO-34 molecular sieve under hydrothermal aging conditions. [8] Ma et al. [9] found that the hydrothermal stability of Cu/SAPO-34 in NH_3-SCR was significantly improved by physically mixing Cu/SAPO-34 with Cu/SSZ-13. This is because both SAPO-34 and SSZ-13 are CHA, and during hydrothermal treatment, SAPO-34 migrates and fills the defect structure generated by dealumination in the SSZ-13 skeleton. These processes can suppress the fracture of Si–O–Al bonds and prevent further damage to the SSZ-13 skeleton. Therefore, the complete SSZ-13 skeleton structure can provide sufficient ion exchange sites for isolated Cu^{2+} species, reduce the migration and aggregation of isolated Cu^{2+} in hydrothermal processes, and thus enable the catalyst to exhibit strong hydrothermal stability.

SSZ-13 is an aluminum silicate CHA type material with high silicon content. Peng et al. [7] prepared a series of Cu/SSZ-13 molecular sieves using a two-step hydrothermal method, with grain sizes ranging from 50 to 300 nm. Compared with other conventional hydrothermal methods, Cu/SSZ-13 zeolite prepared by two-step hydrothermal method has smaller particle size and superior hydrothermal stability. Feng et al. [10] found that Fe/ZSM-5 with a crystal size of approximately 50 nm (cauliflower like) exhibited high NH_3-SCR activity. Therefore, the synthesis of small crystalline catalysts is an important way to improve the hydrothermal stability of NH_3-SCR metal-based zeolites. Christopher et al. [11] modified Cu/SZ-13 with different P contents by adding phosphoric acid through the IWI method. They found that an appropriate amount of P can coordinate with the framework to stabilize the SSZ-13 zeolite framework by inhibiting framework dealumination, ensuring that the separated Cu^{2+} species are better anchored in the zeolite, and ultimately enhancing the hydrothermal stability of Cu/SSZ-13 in NH_3-SCR. However, excessive P can lead to the formation of phosphate species, reducing the concentration of active Cu^{2+} species separated in the zeolite, thereby reducing the NH_3-SCR activity of Cu/SSZ-13 catalyst.

7.2.3 Core–Shell Structure in Copper Based NH_3-SCR Catalysts

For core–shell structure materials, they can show the characteristics of core and shell materials, and have been widely studied in heterogeneous catalysis. In recent years, some researchers have studied the application of core–shell catalysts in NH_3-SCR reactions and found that core–shell catalysts have high activity [12].

Huang et al. [13] successfully prepared a novel core–shell structured nanorod $Cu_mV_nO_y$-NF@Ce-MOF, by in-situ growth of Ce-MOF on copper vanadate precursor nanofibers, as shown in Fig. 7.1. Calcined at 500, 600, and 700 °C to obtain $Cu_2V_2O_7$@CeO_2, $Cu_3(VO_4)_2$@CeO_2, and $Cu_{11}O_2(VO_4)_6$@CeO_2, the CeO_2 shell not only protects the fracture of copper vanadate nanofibers during the calcination process, but also induces the interaction between Ce, Cu, and V, giving them good redox ability, demonstrating their potential as denitrification catalysts. A series of characterizations indicate that the crystallinity of CeO_2 on the surface of $Cu_mV_nO_y$@CeO_2-WO_x decreased significantly, and the acid content on the surface increased significantly. In addition, different calcination temperatures also have an undeniable impact on the surface acid content and redox ability of composite catalytic materials. $Cu_3(VO_4)_2$@CeO_2-WO_x calcined at 600 °C has the maximum total amount of acid, acidity, and reducing ability, and exhibits good catalytic performance in the NH_3-SCR process, which the NOx conversion rate remains above 90% at 230–380 °C.

Fig. 7.1 Schematic diagram of core–shell material preparation ($Cu_mV_nO_y$-NF@Ce-MOF)

Yu et al. [14] prepared a novel CuO@Cu-MOFs core shell material, which is shown in Fig. 7.2. The as-prepared CuO@Cu-MOFs material combines the physical and chemical properties of CuO (high NO adsorption stability and excellent surface reducibility) and $Cu_3(BTC)_2$ (large specific surface area and abundant acidic sites). In addition, the core–shell material contains many Brønsted acid sites and abundant adsorbed NO_x intermediate products, which is conducive to low-temperature NH_3-SCR reaction. The experimental results indicated that CuO@Cu-MOFs has higher low-temperature catalytic activity than pure CuO and $Cu_3(BTC)_2$. Additionally, CuO@Cu-MOFs exhibits high stability in the NH_3-SCR reaction and is a potential low-temperature SCR catalyst.

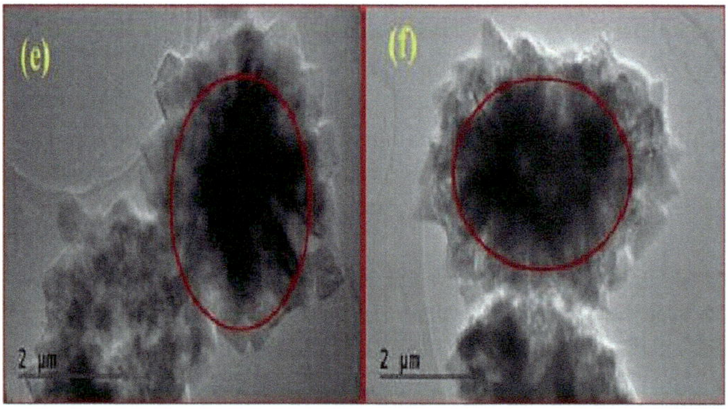

Fig. 7.2 TEM diagram of CuO@Cu-MOFs core–shell material

7.3 Fe-Based SCR Catalysts

7.3.1 Iron Oxide-Based Catalyst

As an environmentally friendly catalyst, Fe_2O_3 has advantages such as high thermal stability, excellent activity at medium to high temperatures, N_2 selectivity, and stable sulfur resistance above 300 °C, thus possessing strong strength in the field of coal-fired flue gas denitrification. The main applications in SCR reactions are α-Fe_2O_3 and γ-Fe_2O_3. [15] As shown in Fig. 7.3, within the temperature range of 150 to 300 °C, the denitrification activity of γ-Fe_2O_3 is significantly higher than α-Fe_2O_3 and maintain a 90% NO conversion rate between 200 and 300 °C, which is contributed to the different adsorption ability of NH_3 over α-Fe_2O_3 and γ-Fe_2O_3 catalysts. γ-Fe_2O_3 is more prone to adsorbing NH_3, while the stability of the adsorbed nitrate species is poor. SCR reactions can be effectively carried out through both the E–R and L–H reaction pathways [16]. Although γ-Fe_2O_3 has better low-temperature activity, but its thermal stability is poor, and it is easily converted into α-Fe_2O_3 at above 320 °C temperature, causing an irreversible decrease in catalyst activity. Qu et al. [17] selected inert Ti^{4+} and Zn^{2+} as substitutes of Fe^{3+} on octahedron vacancy and tetrahedral vacancy in γ-Fe_2O_3, which suppress the conversion of γ-Fe_2O_3 to α-Fe_2O_3. The results showed that the addition of Ti^{4+} did not affect the catalytic activity. After adding Zn^{2+}, the catalyst activity significantly decreased, indicating that Fe^{3+} on tetrahedral vacancies was the main active site in the NH_3-SCR reaction. For Fe based catalysts, activation energy, specific surface area, and morphology are important factors for evaluating their performance. A lower activation energy is beneficial for the progress of SCR reaction. In 2010, Yao et al. [18] calculated that the activation energy of Fe_2O_3 catalyst was 28.32 kJ/mol, significantly lower than that of V_2O_5/TiO_2 (85 kJ/mol) [19], H-ZSM-5 (55 kJ/mol) [20], and MnO_x/TiO_2 (38 kJ/mol) [21].

The Fe_2O_3 catalyst prepared by Wang and Gui's research team [22] has a specific surface area of 90.79 m^2/g, which is significantly larger than the commercial vanadium-based catalyst (83.76 m^2/g). During the SCR reaction process, the increased specific surface area significantly increased the adsorption capacity of gas-phase reaction species on the catalyst surface, promoting the Fe_2O_3 catalyst to achieve a 95% NO conversion rate at 180 °C. Xing et al. [23] synthesized different morphologies using a simple hydrothermal method α-Fe_2O_3 catalyst: nanorods and nanotubes. The difference in catalytic activity between the two catalysts depends on the difference in surface structure. The author found through SEM and HRTEM characterization that α-Fe_2O_3 nanorods mainly expose the (110) crystal surface, resulting in more Fe species on the catalyst surface and higher denitrification performance. Huang et al. [24] also found that the morphology of α-Fe_2O_3 catalyst was regulated, and its activity sequence was hexagonal > nanorods $\approx$ rhombic. This is because the hexagonal shape of (001) is mainly exposed α-Fe_2O_3 has the strongest adsorption capacity for NH_3 species. In addition, the (110) and (001) crystal planes are mainly

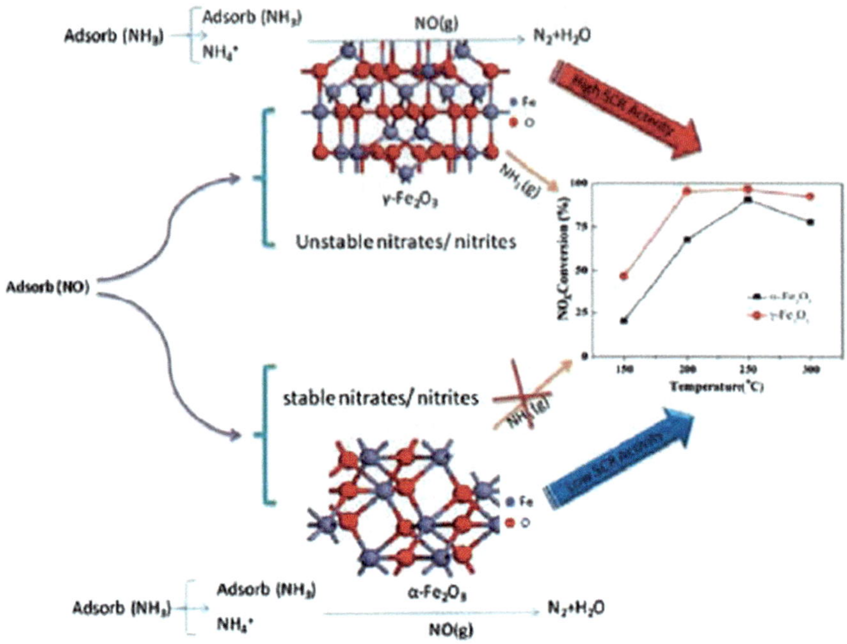

Fig. 7.3 The possible reaction mechanism on α-Fe_2O_3 and γ-Fe_2O_3 catalysts proposed by Liu et al. [16]

exposed the γ-Fe_2O_3 nanorod catalyst exhibits an 80% NO conversion rate and over 98% N_2 selectivity in the temperature range of 200–400 °C. [25]

The above are all SCR catalysts with Fe_2O_3 as a single component. Many research reports suggest that multi-component catalysts can optimize each other's structure and properties, such as obtaining a larger specific surface area, appropriate redox properties, enhanced surface acidity, and thus better catalytic activity and anti H_2O/SO_2 performance. Fe_2O_3 plays different roles in multi-component catalysts.

(1) Fe_2O_3 as bulk dopant

Fe based composite oxide catalysts can effectively regulate the structure of catalysts, optimize their catalytic properties, and improve their resistance to H_2O/SO_2. Therefore, they play an important role in NH_3-SCR. Metal elements such as Ti, Mn, W, V, Cr, and Ce are often incorporated into Fe based catalysts, effectively improving their SCR activity. Zhang et al. [26] prepared a Fe–V–O composite oxide catalyst using a coprecipitation method, which achieved a denitrification efficiency of over 80% in the temperature range of 210–400 °C. Li et al. [27] prepared MnFeOx thin plate catalysts and investigated their activity in NH_3-SCR reactions. The XPS and in situ DRIFTs results showed that the formation of coupled Mn_2O_3 nanocrystals improved its low-temperature activity, and the catalyst achieved a 98% NO conversion rate at 200 °C, 30,000 h^{-1} space velocity, and achieved better SO_2/H_2O resistance and thermal stability.

The preparation methods commonly used to obtain composite oxides include hydrothermal method, coprecipitation method, sol gel method, etc. Liu et al. [28] studied the effects of different precursor salts and preparation methods on the NH_3-SCR activity of Fe_xTiO_y catalyst. It was found that the catalyst prepared by co-precipitation method has significant activity advantages, and compared to $TiCl_4$, using $Ti(SO_4)_2$ as the precursor salt of Ti can improve the catalytic activity and N_2 selectivity of the catalyst.

The problem of SO_2 poisoning is also a major challenge faced by Fe based composite oxide catalysts. Liu et al. [29] investigated the effect of SO_2 on the NH_3-SCR activity of $FeTiO_x$ catalyst at different temperatures. The results showed that the SCR reaction of $FeTiO_x$ at low temperatures was mainly carried out through the L–H mechanism. The coverage of sulfate species on the catalyst surface inhibits the effective adsorption of active species NO_x, hinders the preformation of the L–H reaction pathway, and leads to a decrease in low-temperature activity. When the temperature rises above 250 °C, the author found that the reaction on the sulfated $FeTiO_x$ catalyst mainly follows the E–R mechanism, which is the main reason why the catalyst achieves excellent sulfur resistance performance at high temperatures.

(2) Fe_2O_3 as active component

Researchers often load Fe_2O_3 on the surface of other carriers to improve the denitrification performance of the catalyst. Wang et al. [30] prepared a series of Fe–Mn/Al_2O_3 catalysts, among which 8Fe–8Mn/Al_2O_3 achieved a 99% NO conversion rate at conditions as low as 150 °C. However, Fe based catalysts supported on Al_2O_3 exhibit poor sulfur resistance, which limits their application in NH_3-SCR. TiO_2 is often used as a carrier for catalysts due to its excellent resistance to SO_2 poisoning and high thermal stability. Shu et al. [31] prepared Fe modified Ce/TiO_2 catalysts by impregnation method and investigated the effect of different Fe/Ti ratios on NH_3-SCR activity. When the Fe/Ti molar ratio is 0.2, the catalyst exhibits the best catalytic activity and sulfur resistance performance. Research has shown that the main reason is the addition of Fe, which promotes the increase of Ce^{3+} and active oxygen species on the catalyst surface, which is conducive to the oxidation of NO to NO_2. Fe_2O_3 is often loaded on carbon materials (AC, CNTs, etc.) with large specific surface area and rich pore structure to achieve better denitrification performance. Zhang et al. [32] prepared a series of $MnFeO_x$/CNTs catalysts, which exhibit amorphous structure, stable morphology, high content of surface active oxygen species, and excellent redox ability. They exhibit excellent low-temperature activity in the temperature range of 80–180 °C at a space velocity of 32,000 h^{-1}. Its activity sequence is 12% Mn FeO_x/CNTs > 10% $MnFeO_x$/CNTs > 8% $MnFeO_x$/CNTs > 14% $MnFeO_x$/CNTs.

(3) Fe_2O_3 as a carrier

Using Fe_2O_3 as the carrier of the supported NH_3-SCR catalyst, the active components loaded can be highly dispersed on the surface of Fe_2O_3, and Fe_2O_3 itself has high thermal stability, excellent redox ability, and resistance to SO_2 poisoning.

The synergistic effect between Fe_2O_3 support and active components can effectively improve the denitrification performance of the catalyst. Liu et al. [33] investigated the SCR activity on WO_3/Fe_2O_3 catalysts. The results indicate that the loading of WO_3 provides more Lewis acidic sites for the Fe_2O_3 catalyst, promoting the adsorption and activation of the active species NH_3. In addition, the highly dispersed WO_3 alleviates the crystallization of Fe_2O_3 and optimizes its structure and catalytic properties. When the addition amount of WO_3 is 10%, the catalyst achieves the best NO conversion rate and N_2 selectivity.

7.3.2 *Fe Based Molecular Sieve*

Generally, Fe based molecular sieve catalysts exhibit good catalytic performance in the medium to high temperature range. [34] Therefore, in order to improve the poor low-temperature activity of Fe based molecular sieves, many researchers have attempted to achieve this goal by promoting the "Fast SCR" reaction. Zhang et al. [35] prepared CeO_2-meso-SiO_2@Fe-ZSM-5 core–shell catalyst prioritizes the oxidation of NO on the CeO_2-meso-SiO_2 shell to generate NO_2, followed by a rapid SCR reaction in the Fe-ZSM-5 core structure. This catalyst has higher low-temperature activity than pure Fe-ZSM-5 catalyst. In addition, the activity of the Fe-ZSM-5 catalyst is greatly affected by H_2O, as H_2O is easily adsorbed on the Fe-ZSM-5 catalyst through dipole field interactions, leading to catalyst deactivation. [36] Increasing the Si/Al ratio of Fe-ZSM-5 can improve the hydrophobicity of the catalyst and inhibit the adsorption of H_2O, thereby improving the anti H_2O performance of the catalyst. Du et al. [36] prepared the silicalite-1@Fe-ZSM-5. During the preparation process of the catalyst, Fe^{3+} and Al^{3+} in Fe-ZSM-5 enter the silicalite-1 shell. As a new active site, Fe^{3+} in silicalite-1 enhances its ability to adsorb NO_x, thus showing excellent SCR activity and anti H_2O performance. Sultana et al. [37] added Cu species to the Fe-ZSM-5 catalyst and found that it can enhance the redox ability of the catalyst, improving the low-temperature denitrification efficiency without damaging the high-temperature activity of Fe-ZSM-5.

References

1. Pan W-G, Zhou Y, Guo R-T, et al. Influence of calcination temperature on CeO_2–CuO catalyst for the selective catalytic reduction of NO with NH_3. Environ Prog Sustain Energy. 2014;33(2):385–9.
2. Liu J, Li X, Zhao Q, et al. Mechanistic investigation of the enhanced NH_3-SCR on cobalt-decorated Ce–Ti mixed oxide: In situ FTIR analysis for structure-activity correlation. Appl Catal B Environ. 2017;200:297–308.
3. Wang Q, Xu H, Huang W, et al. Metal organic frameworks-assisted fabrication of CuO/Cu_2O for enhanced selective catalytic reduction of NO_x by NH_3 at low temperatures. J Hazard Mater. 2019;364:499–508.

4. Liu Y-Z, Xu Q-Y, Guo R-T, et al. Enhancement of the activity of Cu/TiO_2 catalyst by Eu modification for selective catalytic reduction of NO_x with NH_3. Environ Sci Pollut Res. 2020;27(22):27663–73.
5. Zheng W, Chen J-L, Guo L, et al. Research progress of hydrothermal stability of metal-based zeolite catalysts in NH_3-SCR reaction. J Fuel Chem Technol. 2020;48(10):1193–210.
6. Wang A, Chen Y, Walter ED, et al. Unraveling the mysterious failure of Cu/SAPO-34 selective catalytic reduction catalysts. Nat Commun. 2019;10(1):1137.
7. Peng C, Liu Z, Horimoto A, et al. Preparation of nanosized SSZ-13 zeolite with enhanced hydrothermal stability by a two-stage synthetic method. Microporous Mesoporous Mater. 2018;255:192–9.
8. Buchholz A, Wang W, Xu M, et al. Thermal stability and dehydroxylation of Brønsted acid sites in silicoaluminophosphates H-SAPO-11, H-SAPO-18, H-SAPO-31, and H-SAPO-34 investigated by multi-nuclear solid-state NMR spectroscopy. Microporous Mesoporous Mater. 2002;56(3):267–78.
9. Ma Y, Zhao H, Zhang C, et al. Enhanced hydrothermal stability of Cu-SSZ-13 by compositing with Cu-SAPO-34 in selective catalytic reduction of nitrogen oxides with ammonia. Catal Today. 2020;355:627–34.
10. Feng B, Wang Z, Sun Y, et al. Size controlled ZSM-5 on the structure and performance of Fe catalyst in the selective catalytic reduction of NO_x with NH_3. Catal Commun. 2016;80:20–3.
11. Paolucci C, Parekh AA, Khurana I, et al. Catalysis in a cage: condition-dependent speciation and dynamics of exchanged Cu cations in SSZ-13 zeolites. J Am Chem Soc. 2016;138(18):6028–48.
12. Pant A, Mackraj I, Govender T. Advances in sepsis diagnosis and management: a paradigm shift towards nanotechnology. J Biomed Sci. 2021;28(1):6.
13. Huang X, Dong F, Zhang G, et al. Modification of composite catalytic material $CumVnO_x$@CeO_2 core–shell nanorods with tungsten for NH_3-SCR. Nanoscale. 2020;12(30):16366–80.
14. Yu Y, Chen C, He C, et al. In situ growth synthesis of CuO@Cu-MOFs core-shell materials as novel low-temperature NH_3-SCR catalysts. ChemCatChem. 2019;11(3):979–84.
15. Shavel A, Liz-Marzán LM. Shape control of iron oxide nanoparticles. Phys Chem Chem Phys. 2009;11(19):3762–6.
16. Liu C, Yang S, Ma L, et al. Comparison on the performance of α-Fe_2O_3 and γ-Fe_2O_3 for selective catalytic reduction of nitrogen oxides with ammonia. Catal Lett. 2013;143(7):697–704.
17. Qu W, Chen Y, Huang Z, et al. Active tetrahedral iron sites of γ-Fe_2O_3 catalyzing NO reduction by NH_3. Environ Sci Technol Lett. 2017;4(6):246–50.
18. Yao G-H, Wang F, Wang X-B, et al. Magnetic field effects on selective catalytic reduction of NO by NH_3 over Fe_2O_3 catalyst in a magnetically fluidized bed. Energy. 2010;35(5):2295–300.
19. Amiridis MD, Wachs IE, Deo G, et al. Reactivity of V_2O_5 catalysts for the selective catalytic reduction of NO by NH_3: influence of Vanadia loading, H_2O, and SO_2. J Catal. 1996;161(1):247–53.
20. Stevenson SA, Vartuli JC, Brooks CF. Kinetics of the selective catalytic reduction of NO over HZSM-5. J Catal. 2000;190(2):228–39.
21. Wu Z, Jiang B, Liu Y, et al. Experimental study on a low-temperature SCR catalyst based on MnO_x/TiO_2 prepared by sol–gel method. J Hazard Mater. 2007;145(3):488–94.
22. Wang X, Gui K. Fe_2O_3 particles as superior catalysts for low temperature selective catalytic reduction of NO with NH_3. J Environ Sci. 2013;25(12):2469–75.
23. Yang X-Y, Li B, Sun L, et al. Effect of surface structure of Fe_2O_3 on the selective catalytic reduction of NO by NH_3. Acta Phys Chim Sin. 2012;28(1):184–8.
24. Huang Z, Du Y, Zhang J, et al. Reduction of NO with NH_3 over ferric oxide nanocrystals: the crystallographic facet-induced catalytic enhancement. Catal Sci Technol. 2018;8(18):4702–8.
25. Mou X, Zhang B, Li Y, et al. Rod-shaped Fe_2O_3 as an efficient catalyst for the selective reduction of nitrogen oxide by ammonia. Angew Chem Int Ed. 2012;51(12):2989–93.
26. Zhang P, Li D. Selective catalytic reduction of NO with NH_3 over iron-vanadium mixed oxide catalyst. Catal Lett. 2014;144(5):959–63.

27. Li Y, Wan Y, Li Y, et al. Low-temperature selective catalytic reduction of NO with NH_3 over Mn_2O_3-doped Fe_2O_3 hexagonal microsheets. ACS Appl Mater Interfaces. 2016;8(8):5224–33.
28. Liu F, He H, Zhang C, et al. Selective catalytic reduction of NO with NH_3 over iron titanate catalyst: catalytic performance and characterization. Appl Catal B Environ. 2010;96(3):408–20.
29. Liu F, He H, Zhang C, et al. Mechanism of the selective catalytic reduction of NO_x with NH_3 over environmental-friendly iron titanate catalyst. Catal Today. 2011;175(1):18–25.
30. Wang X, Wu S, Zou W, et al. Fe-Mn/Al_2O_3 catalysts for low temperature selective catalytic reduction of NO with NH_3. Chinese J Catal. 2016;37(8):1314–23.
31. Shu Y, Sun H, Quan X, et al. Enhancement of catalytic activity over the iron-modified Ce/TiO_2 catalyst for selective catalytic reduction of NO_x with ammonia. J Phys Chem C. 2012;116(48):25319–27.
32. Zhang Y, Zheng Y, Wang X, et al. Preparation of Mn–FeO_x/CNTs catalysts by redox co-precipitation and application in low-temperature NO reduction with NH_3. Catal Commun. 2015;62:57–61.
33. Liu Z, Su H, Chen B, et al. Activity enhancement of WO_3 modified Fe_2O_3 catalyst for the selective catalytic reduction of NO_x by NH_3. Chem Eng J. 2016;299:255–62.
34. Brandenberger S, Kröcher O, Tissler A, et al. The state of the art in selective catalytic reduction of NO_x by ammonia using metal-exchanged zeolite catalysts. Catal Rev. 2008;50(4):492–531.
35. Zhang L, Du T, Qu H, et al. Synthesis of Fe-ZSM-5@Ce/mesoporous-silica and its enhanced activity by sequential reaction process for NH_3-SCR. Chem Eng J. 2017;313:702–10.
36. Du T, Qu H, Liu Q, et al. Synthesis, activity and hydrophobicity of Fe-ZSM-5@silicalite-1 for NH_3-SCR. Chem Eng J. 2015;262:1199–207.
37. Sultana A, Sasaki M, Suzuki K, et al. Tuning the NO_x conversion of Cu-Fe/ZSM-5 catalyst in NH_3-SCR. Catal Commun. 2013;41:21–5.

Chapter 8
Chemical Deactivation and Resistance of Low-Temperature SCR Catalyst

8.1 Introduction

Nitrogen oxide (NO_x) is the umbrella term given to a sequence of active gases, including nitrogen oxide (NO), nitrous oxide (N_2O), and nitrogen dioxide (NO_2), and so on. It is one of the primary atmospheric pollutants formed from thermal power plants and industrial boilers (stationary sources) [1]. Besides the toxicity to human beings directly, it leads to acid rain, photochemical smog and can influence global tropospheric chemistry. A recent investigation indicated that government departments had carried out a full arsenal of activities to control NO_x emissions but yielded no satisfying results [2]. Consequently, confronted with increasingly severe nitrogen oxide pollution, it has become the shared objective to find an efficient and economical treatment technology to reduce NO_x emission.

Among all well-established commercial technologies to eliminate NO_x, selective catalytic reduction (SCR) is a broadly employed technology owing to its superior economy and preferable denitrification efficiency. As the decisive factor of the SCR technology, the performance of the catalysts immediately affects the denitration efficiency of the SCR system [3–6]. To date, abundant studies have been carried out to develop SCR catalysts and yielded plenty of positive results that include precious metals (Pt, Rh, Pd, etc.) and transition metals (Co, Mn, Ce, Ni, Cr, Cu, etc.) [7–17]. V_2O_5/TiO_2 catalyst is a well-established and typical catalyst with the excellent efficiency and commercialized for ages. Though this catalyst shows excellent catalytic activity, there are still many shortcomings related with them. First, vanadium-based oxide catalysts are toxic with a narrow operating window, and generate large amounts of SO_3 and N_2O when the temperature exceeds 350 °C. Secondly, SCR reactors is usually installed behind economizers or in front of air preheater to utilize the heat of the hot exhaust gas., where the catalysts is easily poisoned owing to dust accumulation and SO_2. These poisons has a negative effect on SCR performance and shorten their work lifetime, and may even plug catalysts module or monolith pores, causing to the rise of the backpressure of flue gas and leading an explosion. Thus, non-vanadia

W. Pan and R. Guo, *Low-Temperature Selective Catalytic Reduction Catalysts*,
https://doi.org/10.1007/978-981-97-9298-6_8

catalysts with superior efficiency at low temperature has gradually become a hot research topic in recent years [18–20].

Manganese oxide (MnO_x) catalyst has become an attractive research area in recent years owing to its excellent NO conversion and N_2 selectivity. Similarly, due to its good structural properties and strong ability to interact with other components, ceria is also commonly used as the promoter or carrier of ammonia SCR catalyst. However, manganese-based catalysts and ceria-based catalysts are susceptible to poisoning in the process of application, which has been a thorny problem perplexing many scholars. There are few studies on poisoning of these catalysts compared with vanadium-based. Therefore, this chapter will focus on the poisoning mechanism and anti poisoning research of manganese/cerium based catalysts.

8.2 Deactivation Mechanism of SCR Catalysts by Various Elements

8.2.1 SO_2 and H_2O

In the actual work condition of NH_3-SCR, SO_2 and H_2O exists in the whole reaction process. Scholars generally realize that the existence of SO_2 and H_2O hinders the SCR reaction, and this inhibitory effect is more distinct at low temperature. Therefore, the SO_2 and H_2O deactivation mechanism of SCR catalysts at low-temperature becomes one of the most crucial issues for its commercial-scale application.

(1) **SO_2**

Catalyst deactivation induced by SO_2 is an inevitable problem for the SCR process because of the significant content of SO_2 in the flue gas. Accordingly, many scholars have studied and expounded the SO_2 deactivation mechanism of manganese-based catalysts in detail [21–23]. Park et al. confirmed that SO_2 was easily oxidized to SO_3, then reacted with NH_3 and H_2O to form ammonium sulphate (NH_4HSO_4/$(NH_4)_2SO_4$) [24]. These sulfates were challenging to decompose in the experimental temperature range and ultimately deposited on the active site of the catalyst surface. With the further increase of temperature, $MnSO_4$ would form due to the reaction between SO_2 and the active components Mn, resulting in blockage of the active sites. Pan et al. demonstrated the viewpoint mentioned above through experimental research [25]. The SO_2 deactivation mechanism of cerium based catalysts has also been widely studied. Tang et al. confirmed that the sulfation gradually increased with the increase of temperature, resulting in the state of sulfate in ceria changing from surface sulfate to massive sulfate and massive sulfate. At the same time, the formation of bulk and bulk sulfate is the main reason for the decline of catalytic activity [26].

To gain insight into the SO_2 poison of SCR catalysts, the exploration of the relationship between the inhibition of SO_2 and the reaction temperature in the SCR reaction is essential. Jin et al. investigated the SO_2 deactivation of the Mn–Ce/TiO_2

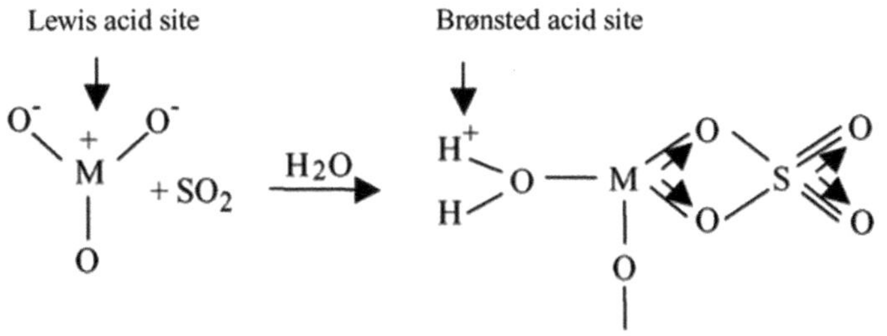

Fig. 8.1 SO_2 poisoning effect of Lewis acid sites on Mn–Ce/TiO_2 catalyst (M stands for Mn–Ce–Ti metal oxide). Reproduced from Ref [27]. Copyright 2010 Elsevier

catalyst and found that the reaction temperature was proportional to the catalyst activity [27]. They also pointed out that the active sites of the catalyst were severely sulfated at 200 °C, which led to the worst irreversible deactivation. For Mn–Ce/TiO_2 catalyst, the inhibition of SO_2 on the Lewis acid sites on is shown in Fig. 8.1. When the SCR reaction was carried out at 200 °C, the adsorbed SO_2 on the surface led to the reduction of initial Lewis acid sites and the increase of Brønsted acid sites. Shen et al. prepared MnO_x CeO_x/ACF catalyst by impregnation method and simulated the SO_2 poisoning of the catalyst [28]. The experimental results manifested that the deactivation degree of SO_2 was closely related to temperature.

From the above study, SO_2 usually acts on the surface of the catalyst in the following three ways to gradually deactivate the catalyst: (1) When the flue gas temperature is lower than 250 °C applying NH_3 as a reducing agent, water vapor in the flue gas and SO_3 formed by the oxidation of SO_2 will continue to react on the catalyst's surface to form ammonium sulphate (NH_4HSO_4/$(NH_4)_2SO_4$). Ammonium sulphate is difficult to decompose in this temperature range and gradually deposits on the surface or internal pores of the catalyst, leading to a gradual decrease in the activity of the catalyst. (2) SO_2 can straightforward to react with the active components in the Mn-based catalyst to form sulphate, which blocks the active site of the catalyst surface. Furthermore, the damage to the catalyst caused by the sulfation would be irreversible. (3) The competitive adsorption of SO_2 can interfere with the intermediate substances forming, resulting in the reduction of NO_x conversion rate. Figures 8.2, 8.3 and 8.4 show the SO_2 deactivation mechanism of manganese-based catalysts.

(2) H_2O

Water appears in the flue gas in water vapor, which usually condenses on the catalyst's surface. Previous literatures demonstrated that H_2O slightly affected the crystal structure, grain size, and specific surface area of manganese-based catalysts. To crown all, water vapor would occupy a series of acid sites and reduce the number of active sites on the catalyst's surface [30, 31].

The adverse effects of H_2O are typically classified into two aspects: reversible and irreversible deactivation. The reversible deactivation refers to the competitive adsorption of H_2O and reaction gas NH_3 on the Lewis acid sites, which will lead

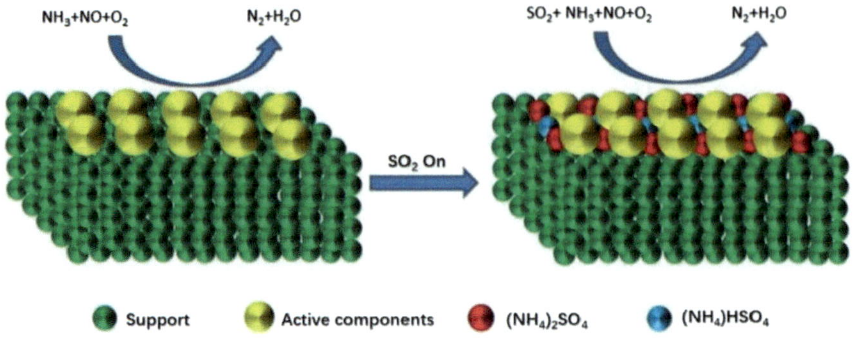

Fig. 8.2 The formation process of $(NH_4)_2SO_4$ and NH_4HSO_4. Reproduced from Ref [29]. Copyright 2018 Elsevier

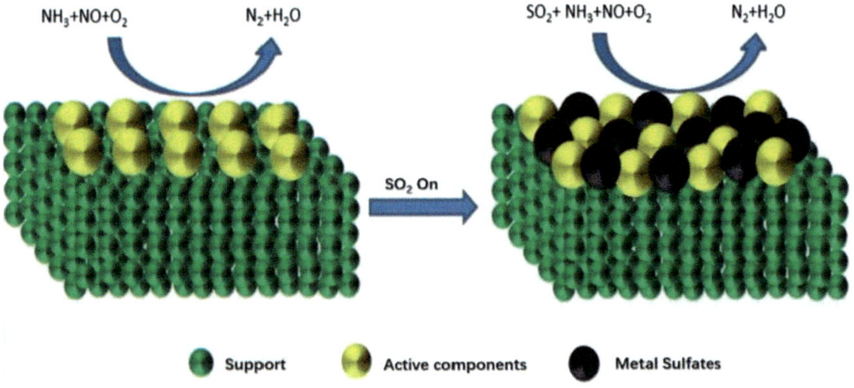

Fig. 8.3 The formation process of sulphate. Reproduced from Ref [29]. Copyright 2018 Elsevier

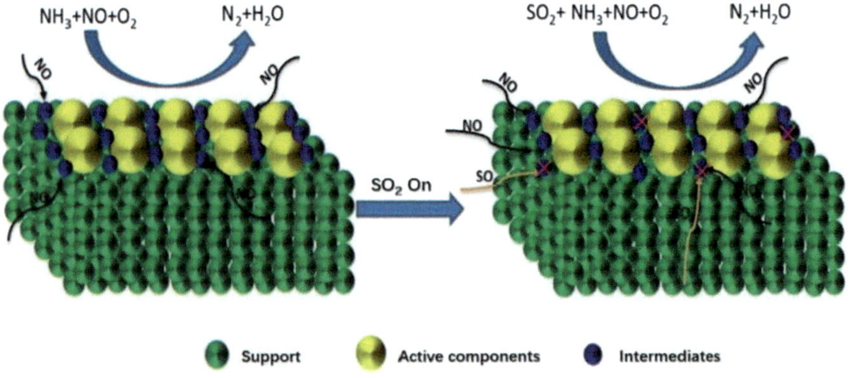

Fig. 8.4 Competitive adsorption of NO and SO_2. Reproduced from Ref [29]. Copyright 2018 Elsevier

to the blockage of the catalyst's surface and the diminishment of the NO conversion rate. Fortunately, as the temperature rises further, the inhibition condition of the catalyst will gradually disappear, and the activity will recover. The experimental data show that the toxic of water is noticeable as the operating temperature is below 200 °C. However, if the reaction temperature rises to more than 200 °C, it is difficult for water vapor to adsorb on the catalyst's surface, and the inhibitory effect is almost negligible. By contrary, the deactivation caused by the dissociative adsorption and decomposition of H_2O on the catalyst's surface is irreversible. The surface hydroxyl (–OH) formed by the decomposition of H_2O on the catalyst surface can only be removed at high temperatures (252–502 °C), and the resulted low-temperature catalyst deactivation is irreversible. Hu et al. demonstrated the inhibition of H_2O on the activity of MnO_2 catalyst, and the denitration rate at 100 °C decreased from 90 to 10% [32]. With the increase of temperature, the inhibition of H_2O on SCR performance continued to weaken, and the denitration rate was unaffected at 300 °C. A similar result had also been reported by Guo et al. [33], who conducted a water vapor effect investigation on SCR performance over CeO_2–CuO/Al_2O_3 catalysts.

In practical application, water and sulfur dioxide generally exist simultaneously. The synergistic poisoning impact of H_2O and SO_2 can aggravate the formation of sulfate and accelerate the deactivation rate of SCR catalyst. Liu et al. studied the inhibitory impact of H_2O and SO_2 on $Fe_{0.75}Mn_{0.25}TiO_x$, and the results exhibited that the influence of H_2O on the SCR activity of the catalyst was slight and reversible [34]. In contrast, the inhibitory effect of SO_2 was more decisive and irreversible, because sulphate was deposited on the catalyst's surface and cut off the path of the SCR reaction, as illustrated in Fig. 8.5. Sun et al. investigated the poisoning effect of SO_2/H_2O on Ce/TiO_2 catalyst. after the introduction of 100 ppm SO_2 and 5% H_2O, the NO_x conversion over Ce/TiO_2 catalyst drops from about 91% to about 75% in 2 h [35].

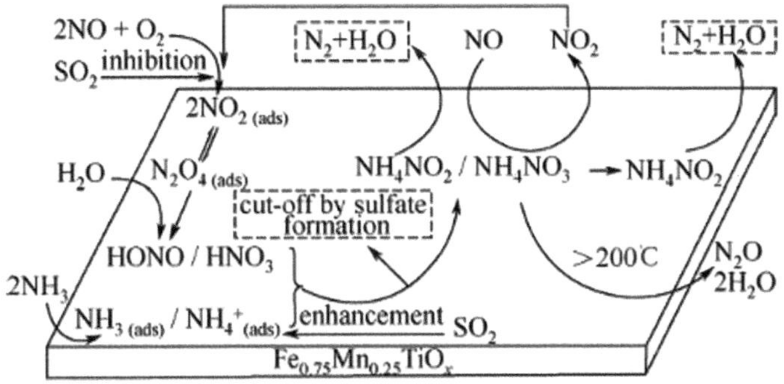

Fig. 8.5 The influence path of SO_2 on NH_3-SCR. Reproduced from Ref [34]. Copyright 2010 Elsevier

8.2.2 Alkali Metals/Alkali-Earth Metals

(1) **Alkali poisoning**

In the particulate matter emitted by the flue gas, the content of alkali metals (the main components are sodium sulphate, anhydrous Glauber's salt, etc.) is relatively high, and its impact is also more significant. SCR catalyst poisoning caused by alkali metals includes two mechanisms: physical and chemical poisoning [36, 37]. The viewpoint of physical poisoning is mainly: alkali metal salt particles are deposited on the catalyst's surface, blocking part of the pores, resulting in a decrease in the specific surface area and hindering the contact between the reaction gas and the active components. The chemical poisoning of alkali metals refers to the reaction of alkali metals with the Bronsted acid sites on the catalyst's surface, thereby reducing the overall acidity and reactivity of the catalyst [38, 39]. At present, most studies believe that the chemical poisoning of alkali metals is the leading cause of catalyst deactivation.

The order of toxicity of different alkali metal elements is as follow: $Cs_2O > Rb_2O > K_2O > Na_2O > Li_2O$. In addition to alkali metal oxides, alkali metal sulfates and chlorides will also lead to the deactivation of the catalyst. In the actual operation of coal-fired power plants, the influence of K in alkali metal elements is the most significant because of its relatively high content. Consequently, the research on the chemical poisoning of catalysts primarily focused on K_2O. Previous studies have shown that when the alkali metal particles are smaller than 100 nm, the deposition effect of alkali metal on the catalyst surface is more pronounced [40]. The aerosol form of alkali metal salts can also block the catalyst pores, cover the active sites on the catalyst surface, and also cause the catalyst to deactivate [38]. The research in the laboratory generally uses the impregnation method to impregnate different precursors (such as K_2CO_3, KNO_3). It is supported on the catalyst and calcined to form K_2O, and then the failure of the catalyst is investigated. Analyzed by DRIFT and other characterization methods, Wei et al. proposed the mechanism of potassium poisoning of the Mn/TiO_2 catalyst [41]: The Potassium species deposited on the surface of the catalyst, leading to a significant decrease of surface area and pore volume, which was the first reason for the deactivation of the catalyst. In addition, the addition of potassium ions reduced the adsorption capacity of ammonia at the Lewis acid sites. It strengthened the adsorption capacity of ammonia at the Brønsted acid site, which was confirmed by the analysis of adsorption energy and the density of states based on the DFT calculation. It indicated that after the SCR catalyst K was poisoned, the change of the Bronsted acid site was the other main reason for the decrease in catalytic activity. Guo et al. investigated the impact of several contents of K on the activity of Mn/TiO_2 catalyst and found that the toxicity to the catalyst increased with the increase of K content [42]. The activity decreased sharply after K poisoning, and the specific surface area and pore volume decreased gradually. Inferred from NH_3-TPD test results, the amount of acid on the catalyst surface significantly reduced after K poisoning. The combination of K^+ with the active acid sites hindered the adsorption of NH_3 on the catalyst was the main reason leading to a remarkable decline in the catalyst's activity. Similar conclusions have been reached in the study of potassium poisoning of Ce-based catalysts. Duan et al. studied the potassium poisoning of the Ce–Ti oxide catalyst used for NH_3-SCR and concluded that the potassium species blocked the mesopore [43]. The reduction of specific surface area is not conducive to the adsorption of reactant species, which is the main cause of catalyst poisoning.

(2) **Alkaline earth metal poisoning**

The alkaline earth metals mainly including Ca and Mg can also cause the deactivation of the catalysts. There are two main ways for alkaline earth metals to deactivate the catalyst. One is similar to alkali metals, calcium will occupy the Bronsted acid sites on the catalyst surface and inhibit the oxidation–reduction ability of the catalyst surface. However, its effect is weaker than that of alkali metals [44]. The other is that alkaline earth metals cause catalyst deactivation through physical action [45]. There are mainly two situations in this inhibition: one is that calcium in the flue gas reacts with SO_2 to form $CaSO_4$ and deposits on the surface of the catalyst and in the pores, resulting in the catalyst deactivation; Another situation is that CaO accumulates in the catalyst pores in the form of free oxides, and then reacts with SO_2 or SO_3, causing the particles to increase and block the micropores, preventing the reactants from contacting with the active units, and deactivating the catalyst. Benson et al. believed that alkaline earth metals were usually covered on the surface of SCR catalysts in the form of $CaSO_4$, $CaCO_3$ and $Ca_3Mg(SiO_4)_2$ [46]. The first two compounds reacted with SO_3 and CO_2 in the flue gas. Shen et al. studied Ca deactivation effect on Mn–CeO_x/Ti-PILC catalyst in the SCR reaction and demonstrated that the quantity of Bronsted acid sites on the catalyst surface decreased due to the coverage of alkaline earth metal calcium [47]. The deactivation degree of the catalyst got more intense with the increase in temperature. They also concluded that the decrease of oxygen adsorbed on the surface was the direct cause of catalyst deactivation. Tang et al. proposed the deactivation of Ce/TiO_2-Ca lies in the following two aspects: (1) the inhibited NH_3 adsorption and NO oxidation (< 200 °C); (2) the greatly suppressed NH_3 adsorption and activation (> 200 °C) [45]. Unfortunately, the in-depth study of manganese based alkaline earth metal poisoning is still very rare, and the focus of future work should pay more attention to this point.

8.2.3 *Heavy Metals*

Researches from various countries have found that a terrific amount of heavy metals such as zinc, lead, and arsenic are present in the fly ash from the combustion of power plants. Heavy metals can deactivate the catalyst for selective catalytic reduction reaction. However, systematic research papers about the mechanism of heavy metal poisoning of Mn-based catalysts and Ce-based catalysts are infrequent. As such, this chapter will elaborate on the deactivation mechanism of several common heavy metal elements.

Zinc poisoning is principally caused by the zinc chloride formed during the reaction between zinc and hydrochloric acid. Because of the lower melting point of zinc chloride, it is more susceptible to osmotic poisoning than potassium chloride. The generated zinc chloride will block the pores of the catalyst, deposit on the surface of the catalyst and hinder the SCR reaction [48]. Su et al. discovered that the NO conversion rate of Mn–Ce/AC catalyst under the inhibition of zinc salt was less than 70%. Zn^{2+} in $ZnCl_2$ and $ZnSO_4$ occupied the acidic sites of the catalyst [49], leading to the reduction of NH_3 chemisorption and the obstacle to the formation of intermediate $-NH_2$. Guo et al. studied the toxic effect of Zn on Ce/TiO_2 catalyst, and they concluded that the decrease of surface acidity and chemisorbed oxygen, and the enlargement of TiO_2 nanoparticles should be the main factors leading to the deactivation [50].

Lead poisoning is also one of the causes of catalyst poisoning. The mechanism of lead poisoning can be divided into the following reasons: (1) The introduction of Pb decreases the specific surface area of the catalyst and hamper the adsorption and mass transfer of gas on the catalyst's surface; (2) The doping of Pb makes the catalyst manganese oxide transform from high dispersion or amorphous morphology into crystal structure; (3) Pb has a pronounced effect on the acidic sites of Mn-based SCR catalyst, which dramatically inhibits the adsorption and activation of reductant. Zhou et al. investigated the influence of PbO inhibition on Mn–Ce/TiO_2 catalyst [51]. The results displayed that PbO crystallized manganese oxides, debased the reducibility of manganese and cerium oxides. Moreover, PbO descended the specific surface area and chemisorbed oxygen content, which caused the decline of SCR performance. The study of Khodayari proposed that the accumulation of lead on the catalyst was mainly due to the combination of lead and acid sites [52], which reduced the conversion rate of nitric oxide and affected the adsorption capacity of ammonia.

Arsenic poisoning can be divided into physical and chemical deactivation. Physical deactivation refers to the diffusion of nitrogen oxides to the surface of the catalyst, resulting in the blockage of pores and the reduction of active sites. Ren et al. studied the impact of As poisoning on Mn–Ce/AC catalysts and demonstrated that the formation of As_2O_5 caused catalyst poisoning [53]. As_2O_3 was oxidized to As_2O_5, covered the catalyst surface, hindered the adsorption and diffusion of reactants on the catalyst surface. Chemical deactivation is the reaction between arsenic and the active components of the catalyst, which weakens the acid sites and redox ability of the catalyst surface. Lange et al. used the sublimation method to deactivate the Titania-supported tungsten and molybdenum SCR catalysts by arsenic [54]. They found arsenic deposited on the catalyst surface to form arsenate, which chemically interacted with the active components, so the reactants could not be adsorbed on the active sites, resulting in its deactivation.

8.3 Deactivation Resistance

In the work condition of NH_3-SCR, SO_2, H_2O, alkali/alkaline earth metals (K, Na, Ca), and heavy metals (As, Pb, etc.) exists in the reaction process. Scholars generally realize that the existence of these poisons hinders the SCR reaction, and this inhibitory effect is more distinct at low temperature. Therefore, poisons tolerance of Mn-based catalysts in low-temperature NH_3-SCR becomes one of the most crucial issues for its commercial-scale application.

8.3.1 Resistance to SO_2 or/and H_2O Poisoning

Ammonia will react with water and NO_x to produce strong adhesive ammonium bisulfate at low temperature. During the long-term operation of the catalyst, ammonium bisulfate will gradually accumulate on the catalyst's surface, covering the active sites of the catalyst and shortening the catalyst service life. In response to different working conditions, one or more active component doping methods are often selected to improve the catalyst activity at low temperature; appropriate additives are selected and added to heighten the water-resistance of the catalyst and inhibit the oxidation activity of SO_2. Table 8.1 lists the common methods used for doping manganese-based SCR catalysts to enhance their sulfur and water resistance. It can be concluded from the table that doping Co, Ni and Ce has an excellent role on improving the water resistance and sulfur resistance of Mn-based catalysts, among which CO has the most significant effect. This strategy is also applicable to Ce-based catalysts, and the tolerance of SO_2 will be significantly improved. Sun et al. prepared Ce/TiO_2 catalyst by citric acid method and investigated the effect of Sm doping on SO_2 tolerance of Ce-based catalyst [33]. The results showed that the addition of Sm could effectively enhance the low-temperature SCR activity and SO_2/H_2O resistance of Ce/TiO_2 catalyst, and the appropriate mole ratio of Sm to Ce was 0.3. Similar conclusions have also been confirmed, Liu et al. demonstrated that Eu modification can also significantly improve the SO_2 tolerance of the Ce-based catalyst.

Previous researches have confirmed that the H_2O inhibitory mechanism on SCR catalysts can be divided into competitive physical adsorption (reversible) and chemisorption (irreversible) [32–34, 66, 67]; and the SO_2 inhibitory mechanism on SCR catalyst is intense and irreversible due to the competitive adsorption of sulfur ammonia nitrogen, the deposition of ammonium sulfate, and the sulfation of active components. Tang et al. used four methods (citric acid method, rheological phase method, low-temperature solid phase method and liquid phase co-precipitation method) to prepare pure MnO_x catalysts [55, 66]. Their low-temperature activity was different, and the ability to resist SO_2 poisoning was far from each other. Similar conclusion was reported by Qi et al. as they investigated the SO_2 resistance of MnO_x–CeO_2 catalyst prepared by different methods [64, 68]. Furthermore, Jiang et al. prepared the MnO_x–TiO_2 catalyst by the sol–gel method, the impregnation

Table 8.1 Water and sulfur resistance of element-doped Mn-based SCR catalyst

Catalysts	Original/Deactivated/Recovered activity (loss of activity)	Poisons	References
MnO_x	98%/90%/98% (8.2%)	H_2O	[55]
MnO_2	92%/70%/– (23.9%)	H_2O	[56]
Mn–Ce–O_x	98.7%/90%/– (8.8%)	H_2O	[57]
Mn–Ni–O_x	100%/94%/– (6%)	H_2O	[58]
Mn–Co–O_x	99%/98%/99% (1%)	H_2O	[59]
MnO_x	96%/46%/46% (52.1%)	SO_2	[25]
Mn–Ce–O_x	99%/89%/97% (10.1%)	SO_2	[60]
Mn–Ni–O_x	100%/82%/98% (18%)	SO_2	[58]
Mn–Co–O_x	99%/96%/99% (3%)	SO_2	[59]
Mn–Ce–Sn–O_x	100%/96%/96% (4%)	SO_2	[61, 62]
MnO_x	98%/73%/90% (25.5%)	$H_2O + SO_2$	[55]
Mn–Cu–O_x	95%/64%/90% (32.6%)	$H_2O + SO_2$	[63]
Mn–Ce–O_x	99%/83%/99% (16.2%)	$H_2O + SO_2$	[64]
Mn–Ce–O_x	100%/78%/– (22%)	$H_2O + SO_2$	[65]
Mn–Co–O_x	99%/89%/96% (10.1%)	$H_2O + SO_2$	[59]
Mn–Co–O_x	100%/86%/93% (14%)	$H_2O + SO_2$	[56]
Mn–Ce–Sn–O_x	100%/70%/90% (30%)	$H_2O + SO_2$	[62]

method and the co-precipitation method respectively [69]. Their low-temperature activity and SO_2 resistance were also different. In recent years, many preparation methods have been extensively applied to prepare Mn-based catalysts, such as co-precipitation method [70, 71], citrate method [72], liquid phase oxidation–reduction method [55, 73], and hydrolysis method [74], combustion pyrolysis method [75, 76], template synthesis method [76–78], etc., have achieved brilliant low-temperature activity and a degree of H_2O and SO_2 tolerance. Non-traditional preparation methods are usually applied to prepare Mn-based catalysts with better resistance to SO_2 and H_2O, such as the $Mn_xCO_{3-x}O_4$ catalyst prepared by template synthesis method [78] the uniform spherical Mn–CeO_x catalyst prepared by one-step hydrolysis [79], Sn–MnO_x–CeO_2 catalyst prepared by ultrasonic-assisted co-precipitation method [79]. A similar conclusion was also confirmed on Ce-based catalysts. Guo et al. fabricate CeO_2–CuO/Al_2O_3 catalysts by three different methods and investigated their activities for selective catalytic reduction (SCR) of NO with NH_3. The catalyst prepared by the singlestep sol–gel (SG) method showed the best SCR activity and resistance to SO_2 and H_2O. It can be seen that the preparation method had a noticeable impact on the catalyst's structure, and the structure has a directly connection to the physical and chemical properties of SCR catalyst. Thus, starting from the perspective of the preparation method and structure of the catalyst, preparing SCR catalyst with a unique structure may be a feasible method to improve its resistance to SO_2/H_2O.

To summarise, the water and sulfur poisoning problem of low-temperature SCR. SCR catalysts should be addressed from the poisoning mechanism and the catalytic reaction path. As previously, water inhibits the activity of low-temperature SCR catalysts primarily due to the competitive adsorption between water and ammonia molecules on the catalyst surface. So it is vital to solving the competitive adsorption of water and ammonia when designing the catalyst from the top layer. For example, improving the anti-water poisoning performance of the low-temperature SCR catalyst in terms of its hydrophilicity, hydrophobicity, specific structure, and doping elements; On the other hand, the main reason for the poisoning effect of sulfur dioxide is that the active center component atoms are sulfated or the ammonium sulfate depositing to occupy the active site. Through the design of materials, the theoretical calculation of DFT, SO_2 screening and steric hindrance can be combined to enhance catalyst resistance to sulfur dioxide poisoning.

8.3.2 Resistance to Alkali/Alkaline Metal Poisoning

As mentioned above, the alkali metal poisoning of the catalyst is chiefly resulted from alkali metal elements occupy or inhibit the surface acid sites (mainly Bronsted acid), which abates the catalyst's ability to absorb and activate NH_3. As such, the acidity of the catalyst surface may be the major factor affecting the resistance to alkali metal poisoning [80–82].

Some researchers have reported that doping rare earth metals and transition metal oxides to improve active components could enhance the alkali metal tolerance of Mn-based catalyst. Li et al. investigated the performance of a Co-modified Ni foam supported Mn-based SCR catalyst concerning alkali metal poisoning resistance in the flue gas [83]. After the modification, the quantities of Brønsted acid and Lewis acid sites of the K-poisoned catalyst's surface were significantly increased. Additionally, the reduction performance of $K_{0.3}Co_{0.5}$/fresh catalyst increased substantially compared with the $K_{0.3}$/fresh catalyst. They extrapolated that the added Co interacted with Mn and Ni on the catalyst's surface and improved its performance. Peng et al. proposed that ceria promoted the potassium tolerance of Mn/TiO_2 catalysts based on experiments and DFT calculations [84]. They concluded that cerium provided additional surface acid sites which were difficult to be neutralized by potassium atoms. Beyond that, ceria exhibited excellent reducibility ensuring the activation of adsorbed NH_3 molecules on the surface. Duan et al. investigated the enhancement of kalium resistance of Ce–Ti oxide catalyst for NH_3-SCR reaction by modification with holmium [43]. The result revealed that the introduction of Ho on Ce–Ti could remarkably improve the adsorption of NH_3 and NO_x species on catalyst surface, accompanied by the intensified reactivity of ad-NH_3 species, which should also administer to improve the K resistance.

Significantly, the alkali metal resistance of the catalyst can be further improved by sulfation treatment of the catalyst carrier (such as ZrO_2, TiO_2). Because the sulfated carrier has intense acidity (or super acidity), which can preferentially capture alkali

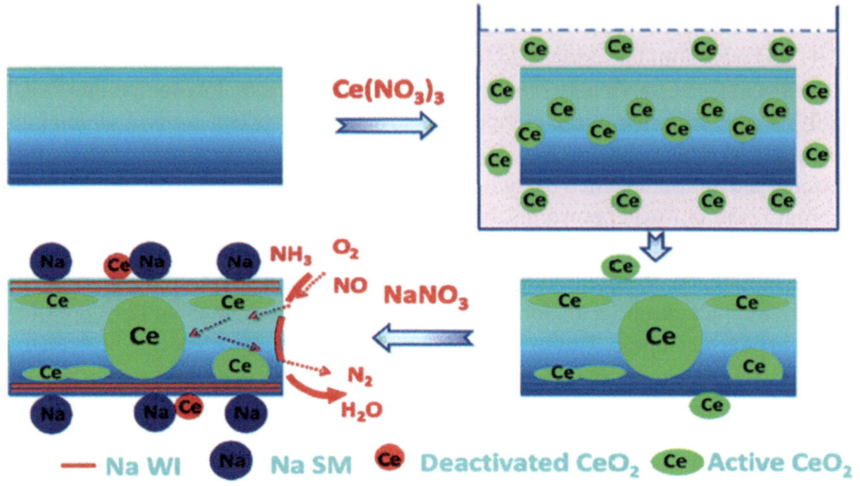

Fig. 8.6 Schematic diagram of the anti-poisoning mechanism of Ce/TNTs. Reproduced from Ref [93]. Copyright 2011 Elsevier

metals, thereby effectively protecting the activity of the component structure not influenced [39, 85, 86]. Sulfated zirconia is an extensively applied solid superacid and has excellent catalytic performance [87, 88]. It has been reported that many active elements loaded on sulfated zirconia show superior SCR catalytic activity, including vanadium [89, 90], copper [32, 90], iron [39], palladium [91, 92], etc. Furthermore, the particular structure and performance of the carrier can also protect the loaded active components from toxic components. Chen et al. loaded the active component CeO_2 in a tube of titanium nanotubes, and the prepared catalyst had excellent alkali metal tolerance [93]. They believed that the unique lumen structure of titanium nanotubes had a superior shell protection effect for the active phase CeO_2 loaded inside, and the interlayer $H_2Ti_{12}O_{25}$ could trap alkali metals (Na^+) through ion exchange, making it free from the influence of alkali metals, as shown in Fig. 8.6.

To sum up, the primary method for SCR catalysts to resist alkali metal poisoning is to strengthen the acidity of the catalyst surface to make it preferentially combine with alkali metals, or use the unique structure of the carrier to protect the active components and avoid the contact of alkali metals with the active phase.

8.3.3 Resistance to Heavy Metal Poisoning

Previous studies have shown that the concentration of Zn in coal-fired fly ash is approximately 3 g/kg [94, 95]. A series of methods have been tested in recent years to heighten the resistance of manganese-based catalysts to zinc poisoning in fly ash. Liu et al. proved that $Al_2(SO_4)_3$ modification significantly enhanced the Zn resistance of Mn/TiO_2 catalyst [96]. The NO conversion of Al-doped catalyst was markedly improved compared with that of undoped catalyst (Fig. 8.7). They concluded that the addition of $Al_2(SO_4)_3$ could effectively abate the crystallinity and produce more acid sites. Further, more Mn_4^+ and chemisorbed oxygen could also lead to the ascendant zinc tolerance of the $MnAl/TiO_2$ catalyst. The doping of $Al_2(SO_4)_3$ enormously improved the zinc tolerance of manganese-based catalysts, and a similar result was also observed on Nb-doped Mn/TiO_2 catalyst [97]. Gong et al. found that multi-layered nanosheets γ-Al_2O_3 with (110) facets could improve the Zn resistance of Ce–Ti catalyst. The result indicated that compared with CeO_2 and TiO_2, the Zn species prefer to absorbed on nanosheets γ-Al_2O_3 with (110) facets, so it can effectively protect CeO_2 from poisoning as a sacrificial agent [98].

Pb is one of the representatives of heavy metals in the municipal solid waste incinerators flue gas and has substantial toxicity to SCR catalyst, which distinctly decreases the activity of the catalyst. Although few reports on enhancing the lead tolerance of Mn-based catalysts, some investigators have made remarkable achievements. Li et al. manifested that the modification of Mn–Ce/Ti catalyst with cobalt (Co) exhibited superior Pb-poisoning resistance [99]. The characterization showed that the doping of Co would increase the dispersion of manganese dioxide. Concurrently, CMCT catalyst had high reducibility and more active sites adsorbing ammonia and NO_x. Jiang et al. found that W modification could significantly improve the lead poisoning tolerance of CeO_2/TiO_2 catalyst [100]. It was confirmed by XPS that the modification of W promoted the transformation of surface Ce^{4+} to Ce^{3+}. Concurrently, additional Ce^{3+} and highly dispersed W species provided more Bronsted

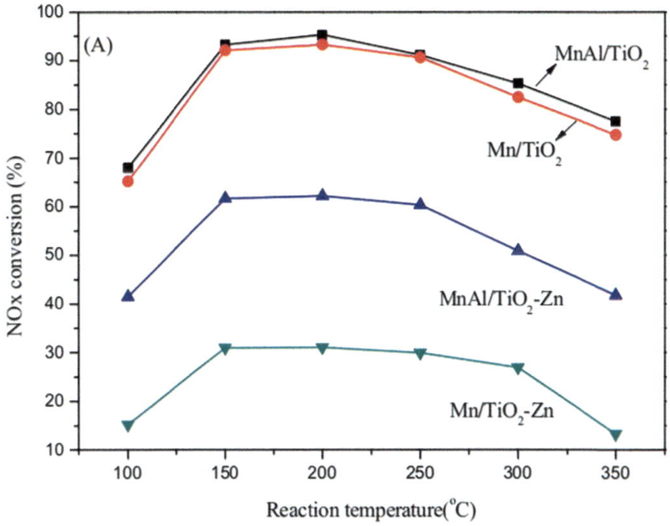

Fig. 8.7 NO_x conversions over the four catalyst samples reaction conditions; [NO] = $[NH_3]$ = 600 ppm, $[O_2] = 5\%$, balance Ar, GHSV = 108,000 h^{-1}. Reproduced from Ref. [96]. Copyright 2017 Elsevier

acid sites for NH_3 adsorption on the CeO_2/TiO_2 catalyst's surface. The methods for improving Pb resistance of Ce-based catalysts can offer valuable information and reference for Mn-based catalysts.

Arsenic causes severe deactivation of SCR catalysts and shorten its working life. Hu et al. identified that SO_4^{2-} had preferable surface acidification performance, and the catalyst with significant SO_4^{2-} support had exceptional As resistance [101]. Thus, we can consider that SO_4^{2-} can be applied as an effective supplement for arsenic resistance of Mn-based catalysts and Ce-based catalysts. Currently, there are few studies on the anti-As poisoning of manganese-based catalysts. As for other SCR catalysts, arsenic still degrades catalytic performance by limiting catalyst surface area, decreasing active sites. Combined with the improved experience of these catalysts [102–104], we infer that the tolerance to As poisoning of Mn-based catalysts and Ce-based can be enhanced by adding active components, doping anti-arsenic additives and carrier modification.

References

1. Lu X, Song C, Chang C, et al. Manganese oxides supported on TiO_2–graphene nanocomposite catalysts for selective catalytic reduction of NO_x with NH_3 at low temperature. Ind Eng Chem Res. 2014;53:11601–10.
2. Thirupathi B, Panagiotis G, et al. Impact of nitrogen oxides on the environment and human health: Mn-based materials for the NO_x abatement. Curr Opin Chem Eng. 2016;13:133–41.
3. Wang Z, Guo R, Shi X, et al. The enhanced performance of Sb-modified Cu/TiO_2 catalyst for selective catalytic reduction of NO_x with NH_3. Appl Surf Sci. 2019;475:334–41.
4. Guo X, Bartholomew C, Hecker W, et al. Effects of sulfate species on V_2O_5/TiO_2 SCR catalysts in coal and biomass-fired systems. Appl Catal B. 2009;92:30–40.
5. Damma D, Ettireddy P, Reddy B, et al. A review of low temperature NH_3-SCR for removal of NO. Catal. 2019;9:349.
6. Zhang P, Guo R, Wu L, et al. The enhancement of NH_3-SCR performance for CeO_2 catalyst by Co pretreatment. Environ Sci Pollut Res. 2020;27:13617–36.
7. Macleod N, Lambert R. Lean NO_x reduction with $CO+H_2$ mixtures over Pt/Al_2O_3 and Pd/Al_2O_3 catalysts. Appl Catal B. 2002;35:269–79.
8. Macleod N, Isaac J, Lambert R. A comparison of sodium-modified Rh/γ-Al_2O_3 and Pd/γ-Al_2O_3 catalysts operated under simulated TWC conditions. Appl Catal B. 2001;33:335–43.
9. Thirupathi B, Smirniotis P. Co-doping a metal (Cr Fe Co Ni Cu Zn Ce and Zr) on Mn/TiO_2 catalyst and its effect on the selective reduction of NO with NH_3 at low-temperatures. Appl Catal B. 2011;110:195–206.
10. Duan C, Guo R, Wu G, et al. Selective catalytic reduction of NO_x by NH_3 over $CeVO_4$-CeO_2 nanocomposite. Environ Sci Pollut Res. 2020;27:22818–28.
11. Liu Y, Xu J, Li H, et al. Rational design and in situ fabrication of $MnO_2@NiCo_2O_4$ nanowire arrays on Ni foam as high-performance monolith de-NO_x catalysts. J Mater Chem A. 2015;3:11543–53.
12. Guo R, Sun X, Liu J, et al. Enhancement of the NH_3-SCR catalytic activity of $MnTiO_x$ catalyst by the introduction of Sb. Appl Catal A. 2018;558:1–8.
13. Zhou C, Zhang Y, Wang X, et al. Influence of the addition of transition metals (Cr Zr Mo) on the properties of MnO_x–FeO_x catalysts for low-temperature selective catalytic reduction of NO_x by Ammonia. J Colloid Interface Sci. 2013;392:319–24.
14. Zhang P, Pan W, Guo R, et al. The Mo modified Ce/TiO_2 catalyst for simultaneous Hg^0 oxidation and NO reduction. J Energy Inst. 2019;92:1313–28.

15. Sun P, Huang S, Guo R, et al. The enhanced SCR performance and SO_2 resistance of Mn/TiO_2 catalyst by the modification with Nb: a mechanistic study. Appl Surf Sci. 2018;447:479–88.
16. Zhou J, Guo R, Zhang X, et al. Cerium oxide-based catalysts for low-temperature selective catalytic reduction of NO_x with NH_3: a review. Energ Fuel. 2021;35:2981–98.
17. Wang Z, Guo R, Shi X, et al. The superior performance of $CoMnO_x$ catalyst with ball-flowerlike structure for low-temperature selective catalytic reduction of NO_x by NH_3. Chem Eng J. 2020;381: 122753.
18. Guo R, Li M, Sun P, et al. Mechanistic investigation of the promotion effect of Bi modification on the NH_3-SCR performance of Ce/TiO_2 catalyst. J Phys Chem C. 2017;121:27535–45.
19. Devaiah D, Dimitrios KP, Thirupathi B, Panagiotis GS. Study of Ce, Sb, and Y exchanged titania nanotubes and superior catalytic performance for the selective catalytic reduction of NO_x. Appl Catal B. 2021;287: 119939.
20. Pappas DK, Thirupathi B, Boolchand P, et al. Novel manganese oxide confined interweaved titania nanotubes for the low-temperature Selective Catalytic Reduction (SCR) of NO_x by NH_3. J Catal. 2016;334:1–13.
21. Guo R, Sun P, Pan W, et al. A highly effective $MnNdO_x$ catalyst for the selective catalytic reduction of NOx with NH_3. Ind Eng Chem Res. 2017;56:12566–77.
22. Chen Q, Guo R, Wang Q, et al. The catalytic performance of Mn/$TiWO_x$ catalyst for selective catalytic reduction of NO_x with NH_3. Fuel. 2016;181:852–8.
23. Zhang S, Zhang B, Liu B, et al. A review of Mn-containing oxide catalysts for low temperature selective catalytic reduction of NO_x with NH_3: reaction mechanism and catalyst deactivation. RSC Adv. 2017;7:26226–42.
24. Park T, Jeong S, Hong S. Selective catalytic reduction of nitrogen oxides with NH_3 over natural manganese ore at low temperature. Ind Eng Chem Res. 2001;40:4491–5.
25. Pan S, Luo H, Li L, et al. H_2O and SO_2 deactivation mechanism of MnO_x/MWCNTs for low-temperature SCR of NO_x with NH_3. J Mol Catal A Chem. 2013;377:154–61.
26. Tang C, Zhang H, Dong L. Ceria-based catalysts for low-temperature selective catalytic reduction of NO with NH_3. Catal Sci Technol. 2016;6:1248–64.
27. Jin R, Liu Y, Wu Z, et al. Relationship between SO_2 poisoning effects and reaction temperature for selective catalytic reduction of NO over Mn–Ce/TiO_2 catalyst. Catal Today. 2010;153:84–9.
28. Shen B, Liu T, Yang T. Deactivation by SO_2 of transition metal oxides modified low-temperature SCR catalyst for NO_x reduction with NH_3. Micropor Mesopor Mater. 2009;8:2204–9.
29. Liu J, Guo R, Li M, et al. Enhancement of the SO_2 resistance of Mn/TiO_2 SCR catalyst by Eu modification: a mechanism study. Fuel. 2018;223:385–93.
30. Lu X, Song C, Jia S, et al. Low-temperature selective catalytic reduction of NO_x with NH_3 over cerium and manganese oxides supported on TiO_2-graphene. Chem Eng J. 2015;260:776–84.
31. Xu G, Guo X, Cheng X, et al. A review of Mn-based catalysts for low-temperature NH_3-SCR: NO_x removal and H_2O/SO_2 resistance. Nanoscale. 2021;13:7052–80.
32. Hu P, Huang Z, Hua W. Effect of H_2O on catalytic performance of manganese oxides in NO reduction by NH_3. Appl Catal A. 2012;437–438:139–48.
33. Guo R, Zhen W, Pan W, et al. Low-temperature selective catalytic reduction of NO on CeO_2–CuO/Al_2O_3 catalysts prepared by different methods. Environ Technol. 2014;35:1766–72.
34. Liu F, He H. Selective catalytic reduction of NO with NH_3 over manganese substituted iron titanate catalyst: reaction mechanism and H_2O/SO_2 inhibition mechanism study. Catal Today. 2010;153:70–6.
35. Sun P, Guo R, Liu S, et al. Enhancement of the low-temperature activity of Ce/TiO_2 catalyst by Sm modification for selective catalytic reduction of NO_x with NH_3. Mol Catal. 2017;433:224–34.
36. Shi Y, Yi H, Gao F. Facile synthesis of hollow nanotube $MnCoO_x$ catalyst with superior resistance to SO_2 and alkali metal poisons for NH_3-SCR removal of NO_x. Sep Purif Technol. 2021;265: 118517.

37. Wang S, Guo R, Pan W, et al. The deactivation of Ce/TiO_2 catalyst for NH_3-SCR reaction by alkali metals: TPD and DRIFT studies. Catal Commun. 2017;89:143–7.
38. Zheng Y, Jensen AD, Johnsson JE, et al. Deactivation of V_2O_5-WO_3-TiO_2 SCR catalyst at biomass fired power plants: elucidation of mechanisms by lab-and pilot-scale experiments. Appl Catal B. 2008;83:186–94.
39. Kustov A, Rasmussen SB, Fehrmann R, et al. Activity and deactivation of sulphated TiO_2-and ZrO_2-based V Cu and Fe oxide catalysts for NO abatement in alkali containing flue gases. Appl Catal B. 2007;76:9–14.
40. Kling Å, Andersson C, Myringer Å, et al. Alkali deactivation of high-dust SCR catalysts used for NO_x reduction exposed to flue gas from 100 MW-scale biofuel and peat fired boilers: influence of flue gas composition. Appl Catal B. 2007;69:240–51.
41. Wei L, Cui S, Guo H, et al. The effect of alkali metal over Mn/TiO_2 for low-temperature SCR of NO with NH_3 through Drift and DFT. Comput Mater Sci. 2018;144:216–22.
42. Guo R, Wang S, Pan W, et al. Different poisoning effects of K and Mg on the Mn/TiO_2 catalyst for selective catalytic reduction of NO_x with NH_3: a mechanistic study. J Phys Chem C. 2017;121:7881–91.
43. Duan C, Guo R, Liu Y, et al. Enhancement of kalium resistance of Ce–Ti oxide catalyst for NH_3-SCR reaction by modification with holmium. J Rare Earths. 2020;40:49–56.
44. Tang F, Xu B, Shi H, et al. The poisoning effect of Na^+ and Ca^{2+} ions doped on the V_2O_5/TiO_2 catalysts for selective catalytic reduction of NO by NH_3. Appl Catal B. 2010;94:71–6.
45. Liu S, Guo R, Wang S. Deactivation mechanism of Ca on Ce/TiO_2 catalyst for selective catalytic reduction of NO_x with NH_3. J Taiwan Inst Chem Eng. 2017;78:290–8.
46. Benson SA, Laumb JD, Crocker CR, et al. SCR catalyst performance in flue gases derived from subbituminous and lignite coals. Fuel Process Technol. 2005;86:577–613.
47. Shen B, Deng L, Chen J. Effect of K and Ca on catalytic activity of Mn–CeO_x/Ti-PILC. Front Environ Sci Eng. 2013;7:512–7.
48. Moradi F, Brandin J, Sohrabi M, et al. Deactivation of oxidation and SCR catalysts used in flue gas cleaning by exposure to aerosols of high-and low melting point salts potassium salts and zinc chloride. Appl Catal B. 2003;46:65–76.
49. Su Z, Ren S, Yang J. Poisoning effect comparison of $ZnCl_2$ and $ZnSO_4$ on Mn–Ce/AC catalyst for low-temperature SCR of NO. Chem Select. 2020;5:9226–34.
50. Guo R, Lu C, Pan W. A comparative study of the poisoning effect of Zn and Pb on Ce/TiO_2 catalyst for low temperature selective catalytic reduction of NO with NH_3. Catal Commun. 2015;59:136–9.
51. Zhou L, Li C, Zhao L. The poisoning effect of PbO on Mn–Ce/TiO_2 catalyst for selective catalytic reduction of NO with NH_3 at low temperature. Appl Surf Sci. 2016;389:532–9.
52. Khodayari R, Odenbrand C. Deactivating effects of lead on the selective catalytic reduction of nitric oxide with ammonia over a V_2O_5/WO_3/TiO_2 catalyst for waste incineration applications. Ind Eng Chem Res. 1998;37:1196–202.
53. Ren S, Li S, Su Z. Poisoning effects of KCl and As_2O_3 on selective catalytic reduction of NO with NH_3 over Mn–Ce/AC catalysts at low temperature. Chem Eng J. 2018;351:540–7.
54. Lange C, Schmelz H, Knozinger H. Infrared-spectroscopic investigations of selective reduction catalysts poisoned with arsenic oxide. Appl Catal B. 1996;8:245–65.
55. Tang X, Hao J, Xu W, et al. Low temperature selective catalytic reduction of NO_x with NH_3 over amorphous MnO_x catalysts prepared by three methods. Catal Commun. 2007;8:329–34.
56. Hossein C, Wu R, Zhang N, et al. Enhancement of low-temperature NH_3-SCR catalytic activity and H_2O resistance ability over MnO_x/TiO_2 catalyst by expanded graphite. Catal Lett. 2020;150:2688–94.
57. Ni S, Tang X, Yi H. Novel Mn–Ce bi-oxides loaded on 3D monolithic nickel foam for low-temperature NH_3-SCR de-NO_x: preparation optimization and reaction mechanism. J Rare Earths. 2020 in press.
58. Wan Y, Zhao W, Yu T, et al. Ni-Mn bi-metal oxide catalysts for the low temperature SCR removal of NO with NH_3. Appl Catal B. 2014;148–9:114–22.

59. Lei Z, Shi L, Lei H, et al. Rational design of high-performance deNO_x catalysts based on $Mn_xCo_{3x}O_4$ nanocages derived from metal–organic frameworks. ACS Catal. 2014;4:1753–63.
60. Chen Z, Wang F, Li H, et al. Low-temperature selective catalytic reduction of NO_x with NH_3 over Fe–Mn mixed-oxide catalysts containing $Fe_3Mn_3O_8$ phase. Ind Eng Chem Res. 2012;51:202–12.
61. Chang H, Li J, Yuan J, et al. Ge Mn-doped CeO_2–WO_3 catalysts for NH_3–SCR of NO_x: effects of SO_2 and H_2 regeneration. Catal Today. 2013;201:139–44.
62. Chang H, Li J, Chen X. Effect of Sn on MnO_x–CeO_2 catalyst for SCR of NO_x by ammonia: enhancement of activity and remarkable resistance to SO_2. Catal Commun. 2012;27:54–7.
63. Kang M, Park ED, Kim JM, et al. Cu-Mn mixed oxides for low temperature NO reduction with NH_3. Catal Today. 2006;111:236–41.
64. Qi G, Yang R, Chang R. MnO_x-CeO_2 mixed oxides prepared by co-precipitation for selective catalytic reduction of NO with NH_3 at low temperatures. Appl Catal B. 2004;51:93–106.
65. Liu Z, Yang Y, Zhang S, et al. Selective catalytic reduction of NO_x with NH_3 over Mn–Ce mixed oxide catalyst at low temperatures. Catal Today. 2013;216:76–81.
66. Tang X, Hao J, Xu W, et al. Nano-$MnO_{(x)}$ catalyst for the selective catalytic reduction of NO by NH_3 in low-temperature. Environ Sci. 2007;28:289–94.
67. Liu F, He H. Structure–activity relationship of iron titanate catalysts in the selective catalytic reduction of NO_x with NH_3. J Phys Chem C. 2010;114:16929–36.
68. Qi G, Yang R. Performance and kinetics study for low-temperature SCR of NO with NH_3 over MnO_x–CeO_2 catalyst. J Catal. 2003;217:434–41.
69. Jiang B, Liu Y, Wu Z. Low-temperature selective catalytic reduction of NO on MnO_x/TiO_2 prepared by different methods. J Hazard Mater. 2009;162:1249–54.
70. Zhang L, Cui S, Guo H, et al. Effect of cerium on manganese base catalysts by co-precipitation for low-temperature SCR of NO with NH_3. Key Eng Mater. 2015;633:121–4.
71. Kang M, Park ED, Kim JM, et al. Cu–Mn mixed oxides for low temperature NO reduction with NH_3. Catal Today. 2006;111:236–41.
72. Guo R, Zhen W, Pan W, et al. Effect of Cu doping on the SCR activity of CeO_2 catalyst prepared by citric acid method. J Ind Eng Chem. 2014;20:1577–80.
73. Sun X, Guo R, Liu J, et al. The enhanced SCR performance of Mn/TiO_2 catalyst by Mo modification: identification of the promotion mechanism. Int J Hydrog Energy. 2018;43:16038–48.
74. Shen B, Wang F, Liu T. Homogeneous MnO_x–CeO_2 pellets prepared by a one-step hydrolysis process for low-temperature NH_3-SCR. Powder Technol. 2014;253:152–7.
75. Andreoli S, Deorsola F, Pirone R. MnO_x-CeO_2 catalysts synthesized by solution combustion synthesis for the low-temperature NH_3-SCR. Catal Today. 2015;253:199–206.
76. Liu Z, Yi Y, Zhang S, et al. Selective catalytic reduction of NO_x with NH_3 over Mn–Ce mixed oxide catalyst at low temperatures. Catal Today. 2013;216:76–81.
77. Qiu M, Zhan S, Yu H, et al. Low-temperature selective catalytic reduction of NO with NH_3 over ordered mesoporous $Mn_xCo_{3-x}O_4$ catalyst. Catal Commun. 2015;62:107–11.
78. Zhang L, Shi L, Huang L, et al. Rational design of high-performance deNO_x catalysts based on $Mn_xCo_{3-x}O_4$ nanocages derived from metal–organic frameworks. ACS Catal. 2014;4:1753–63.
79. Chang H, Chen X, Li J, et al. Improvement of activity and SO_2 tolerance of Sn-modified MnO_x–CeO_2 catalysts for NH_3-SCR at low temperatures. Environ Sci Technol. 2013;47:5294–301.
80. Duan C, Guo R, Liu Y, et al. Enhancement of Kalium resistance of Ce–Ti oxide catalyst for NH_3-SCR reaction by modification with holmium. J Rare Earths. 2022;40:49–56.
81. Guo D, Guo R, Duan C, et al. The enhanced K resistance of Cu-SSZ-13 catalyst for NH_3-SCR reaction by the modification with Ce. Mol Catal. 2021;502: 111392.
82. Yang N, Guo R, Pan W, et al. The promotion effect of Sb on the Na resistance of Mn/TiO_2 catalyst for selective catalytic reduction of NO with NH_3. Fuel. 2016;169:87–92.

83. Li G, Zhu B, Sun Y, et al. Study of the alkali metal poisoning resistance of a Co-modified Mn/Ni foam catalyst in low-temperature flue gas SCR deNO$_x$. J Mater Sci. 2018;53:9674–89.
84. Peng Y, Li J, Si W. Ceria promotion on the potassium resistance of MnO$_x$/TiO$_2$ SCR catalysts: an experimental and DFT study. Chem Eng J. 2015;269:44–50.
85. Putluru SS, Riisager A, Fehrmann R. Alkali resistant Cu/zeolite deNO$_x$ catalysts for flue gas cleaning in biomass fired applications. Appl Catal B. 2011;101:183–8.
86. Putluru SS, Kristensen SB, Due HJ, et al. Alternative alkali resistant deNO$_x$ catalysts. Catal Today. 2012;184:192–6.
87. Heshmatpour F, Aghakhanpour RB. Synthesis and characterization of superfine pure tetragonal nanocrystalline sulfated zirconia powder by a non-alkoxide sol–gel route. Adv Powder Technol. 2012;23:80–7.
88. Yang W, Zhang R, Chen B, et al. New aspects on the mechanism of C_3H_6 selective catalytic reduction of NO in the presence of O_2 over LaFe$_{1-x}$(Cu Pd)$_x$O$_{3-\delta}$ perovskites. Environ Sci Technol. 2012;46:11280–8.
89. Kustov A, Kustova MY, Fehrmann R, et al. Vanadia on sulphated-ZrO$_2$ a promising catalyst for NO abatement with ammonia in alkali containing flue gases. Appl Catal B. 2005;58:97–104.
90. Due HJ, Boghosian S, Kustov A, et al. Vanadia-based SCR catalysts supported on tungstated and sulfated zirconia: influence of doping with potassium. J Catal. 2007;251:459–73.
91. Bautista P, Faraldos M, Yates M, et al. Influence of sulphate doping on Pd/zirconia based catalysts for the selective catalytic reduction of nitrogen oxides with methane. Appl Catal B. 2007;71:254–61.
92. Mirkelamoglu B, Liu M, Ozkan US. Dual-catalyst after treatment of lean-burn engine exhaust. Catal Today. 2010;151:386–94.
93. Chen X, Wang H, Wu Z, et al. Novel $H_2Ti_{12}O_{25}$-confined CeO$_2$ catalyst with remarkable resistance to alkali poisoning based on the "shell protection effect." J Phys Chem C. 2011;115:17479–84.
94. Jegadeesan G, Al AS, Pinto P. Influence of trace metal distribution on its leachability from coal fly ash. Fuel. 2008;87:1887–93.
95. Xiao W, Wang W, Ye T, et al. A study on the chemical and mineralogical characterization of MSWI fly ash using a sequential extraction procedure. J Hazard Mater. 2006;134:197–201.
96. Liu S, Guo R, Peng S, et al. The enhancement of Zn resistance of Mn/TiO$_2$ catalyst for NH$_3$-SCR reaction by the modification with $Al_2(SO_4)_3$. J Taiwan Inst Chem E. 2017;78:370–7.
97. Li W, Guo R, Wang S, et al. The enhanced Zn resistance of Mn/TiO$_2$ catalyst for NH$_3$-SCR reaction by the modification with Nb. Fuel Process Technol. 2016;154:235–42.
98. Gong P, He F, Xie J, et al. Effect of multi-layered nanosheets γ-Al$_2$O$_3$ with (110) facets on zinc deactivation resistance for Ce–Ti SCR catalyst. J Energy Inst. 2020;101:221–32.
99. Li G, Mao D, Chao M. Significantly enhanced Pb resistance of a Co-modified Mn–Ce-Ox/TiO$_2$ catalyst for low-temperature NH$_3$-SCR of NO$_x$. Catal Sci Technol. 2020;10:6368–77.
100. Jiang Y, Yang L, Han D, et al. The enhanced Pb resistance of CeO$_2$/TiO$_2$ catalyst for selective catalytic reduction of NO with NH$_3$ by the modification with W. Mol Catal. 2021;514: 111839.
101. Hu W, Gao X, Deng Y, et al. Deactivation mechanism of arsenic and resistance effect of SO_4^{2-} on commercial catalysts for selective catalytic reduction of NO$_x$ with NH$_3$. Chem Eng J. 2016;293:118–28.
102. Li X, Chen J, Lu C, et al. Performance of Mo modified γ-Fe$_2$O$_3$ catalyst for selective catalytic reduction of NO$_x$ with ammonia: presence of arsenic in flue gas. Fuel. 2021;294: 120552.
103. Jiang S, Li T, Zheng J, et al. Unveiling the remarkable arsenic resistance origin of alumina promoted cerium-tungsten catalysts for NH$_3$-SCR. Environ Sci Technol. 2020;54:14740–9.
104. Wang N, Ye C, Xie H, et al. Fe$_2$O$_3$ enhanced high-temperature arsenic resistance of CeO$_2$-La$_2$O$_3$/TiO$_2$ catalyst for selective catalytic reduction of NO$_x$ with NH$_3$. RSC Adv. 2021;11:9395–402.

GPSR Compliance

The European Union's (EU) General Product Safety Regulation (GPSR) is a set of rules that requires consumer products to be safe and our obligations to ensure this.

If you have any concerns about our products, you can contact us on ProductSafety@springernature.com

In case Publisher is established outside the EU, the EU authorized representative is:

Springer Nature Customer Service Center GmbH
Europaplatz 3
69115 Heidelberg, Germany

Batch number: 10726135

Printed by Printforce, the Netherlands